Behaviour Patterns
of
Blood-sucking Flies

Behaviour Patterns
of
Blood-sucking Flies

E. C. MUIRHEAD-THOMSON

Department of Zoology, Royal Holloway College, Englefield Green, Surrey

PERGAMON PRESS

OXFORD · NEW YORK · TORONTO · SYDNEY · PARIS · FRANKFURT

U.K.	Pergamon Press Ltd., Headington Hill Hall, Oxford OX3 0BW, England
U.S.A.	Pergamon Press Inc., Maxwell House, Fairview Park, Elmsford, New York 10523, U.S.A.
CANADA	Pergamon Press Canada Ltd., Suite 104, 150 Consumers Road, Willowdale, Ontario M2J 1P9, Canada
AUSTRALIA	Pergamon Press (Aust.) Pty. Ltd., P.O. Box 544, Potts Point, N.S.W. 2011, Australia
FRANCE	Pergamon Press SARL, 24 rue des Ecoles, 75240 Paris, Cedex 05, France
FEDERAL REPUBLIC OF GERMANY	Pergamon Press GmbH, 6242 Kronberg-Taunus, Hammerweg 6, Federal Republic of Germany

First edition 1982

Library of Congress Cataloging in Publication Data

Muirhead-Thomson, R. C.
Behaviour patterns of blood-sucking flies.
Bibliography: p.
Includes index.
1. Diptera—Behavior. 2. Bloodsucking animals—Behavior. 3. Entomology—Field work. I. Title.
QL531.M84 1982   614.4′322   81-21019
AACR2

British Library Cataloguing in Publication Data

Behaviour patterns of blood-sucking flies.
1. Flies   2. Parasitic insects
I. Muirhead-Thomson, E. C.
595.77   QL531
ISBN 0-08-025497-7

In order to make this volume available as economically and as rapidly as possible the author's typescript has been reproduced in its original form. This method unfortunately has its typographical limitations but it is hoped that they in no way distract the reader.

Printed in Great Britain by A. Wheaton & Co. Ltd., Exeter

Preface

This book is the outcome of a three-months seminar on "Behavioural Aspects of Sampling Insects of Medical Importance", given at the University of California, Berkeley, in 1977. That course in turn was based on my earlier book on "Ecology of Insect Vector Populations", published in 1969. The many advances in all branches of this subject in the last ten years, and in particular in the three years since the Seminar, have necessitated a great deal of revision, and the inclusion of as many relevant publications as possible up to 1980.

Progress in this wide field of study does not advance steadily on a broad front. Many fruitful research project come to an untimely end for a variety of reasons, and there may be long gaps before the same subject, or the same approach, is once more resumed and revitalized. For this reason several of the outstanding investigations in the 'sixties or earlier contain material and techniques which are equally relevant today, and therefore continue to justify their inclusion alongside the most recent advances, until such time as they in turn need revision in the light of new contributions to that particular field.

Acknowledgements

Grateful acknowledgement is made to the Editors of the following journals for permission to reproduce Figures from publications in their journals:

Bulletin of Entomological Research: Figs. 2.11, 3.4, 3.6–3.10, 3.12, 3.13, 4.4, 4.5, 4.6, 5.3–5.12, 6.3, 6.5, 7.18, 7.19, 8.2, 9.1.
Annals of Tropical Medicine and Parasitology: Figs. 6.1, 7.12–7.15.
Transactions of the Royal Society of Tropical Medicine and Hygiene: Figs. 9.3–9.5.
Journal of Medical Entomology: Figs. 3.1, 3.11, 8.4–8.7, 9.6.
American Journal of Tropical Medicine: Figs. 3.6, 3.8, 4.2, 4.3.
Canadian Journal of Zoology: Figs. 7.1, 7.2, 74.
Journal of Animal Ecology: Figs. 2.12, 5.17.
Annals of the Entomological Society of America: Table 3.1; Figs. 3.2, 3.4, 6.6.
Ecological Entomology: Figs. 5.13, 5.14.
Journal of Economic Entomology: Fig. 6.7.
Entomologists Monthly Magazine: Figs. 7.6, 7.7.
Tropenmedizin und Parasitologie: Figs. 7.9–7.11.
Canadian Entomologist: Fig. 7.5
Entomologia Experimenta et Applicata: Fig. 3.5
Mosquito News: Figs. 4.1, 8.3.
Acta Tropica: Figs. 5.1, 6.4.
Entomological Society of Canada (and Managing Editor of the Proceedings of 10th International Congress on Entomology): Fig. 7.3.
Journal of Applied Ecology: Figs. 8.1, 9.2.
Parassitologia: Fig. 2.8.
ORSTOM: Figs. 5.15, 5.16, 7.8, 7.16, 7.17.
INSERM: Figs. 9.7–9.12 (from INSERM Monograph No. 37).

Contents

CHAPTER 1

Introduction

The two-winged blood-sucking flies which form the subject of this book include some of the most important insect vectors of human disease in the world, to say nothing of their role in transmission of various parasites and pathogens affecting domestic animals and livestock. Mosquitoes themselves include the major carriers of malaria and filariasis, as well as yellow fever, dengue and an imposing list of other viral diseases. Tsetse flies are vectors of various trypanosomes affecting man and livestock in Africa, and play a major role in public health, in human settlement and in land development in that continent. The blackfly vectors of human onchocerciasis are now the focus of particular attention in the World Health Organization Onchocerciasis Control Programme (OCP) in seven adjoining countries of West Africa. The phlebotomine sandflies, long known as vectors of Kala Azar in countries of the East, are still being intensively studied as vectors of various forms of leishmaniasis in Central and South America, as well as in several Mediterranean countries of Europe. In addition to this public health role as transmitters of disease-producing agents or pathogens, some of these insects, not directly concerned with human or animal disease, are serious biting pests of man and livestock in several countries, both within and outside the tropics. For example, blackflies are notorious biting pests in Canada, while biting midges dominate many parts of the Caribbean and the adjacent Gulf States. Such insects provide a serious problem by interfering with outdoor recreation and leisure, and at times their attacks may make conditions intolerable for the unprotected human.

Over the years, intensive studies in all countries affected by biting flies and by the diseases they transmit, have provided a huge fund of knowledge. New information continually adds to this, and it might appear from the sheer bulk of publications that our knowledge about the behaviour and ecology of these insects was by now pretty exhaustive. However, this is very far from being the case. Many different species of blood-sucking fly have been incriminated as disease vectors, and each of these insects has distinctive habits and special relations with its environment, presenting a vast range of ecological situations for investigation. As a result, with a few exceptions, a great deal of recorded information is very incomplete, penetrating advances in one direction being offset by failure or setbacks in others.

The status of each different species of biting fly as vector of disease depends on many different facets of its ecology and behaviour, such as factors determining its abundance, its association with man, its dependence on various climatic conditions, as well as its longevity, distribution and

1

flight range. Each of these aspects requires a different approach and different study methods, and such methods have to be adapted to meet the
special situation presented by each species of vector. Many of these two-
winged blood-sucking flies are small, elusive, and perhaps in addition only
active at night. Consequently many ideas about their movements, flight and
feeding habits can seldom be based on continuous direct observation after
the manner of the classic studies of Fabre, but have to be inferred by
indirect means involving the sampling of the population by means of various
capture or trapping methods. The interpretation of such capture or sampling
data therefore plays a vital part in providing present knowledge about the
ecology and behaviour of these insects. It therefore becomes increasingly
important to comprehend not so much the reported results of such investigations, but the means whereby such information was obtained.

These insects may be captured or trapped by a wide range of different
techniques, some of which depend on attractants such as animal bait, light
traps and CO_2, while others are classed as 'non-attractant' and are designed
to capture the insect in a completely unbiased and objective manner. While
the advantages of each of these many different sampling techniques have
been established for some time, there has been increasing emphasis in recent
years about the disadvantages and limitations of even the most 'efficient'
trapping and capture methods. Different devices tend to sample not only
different species disproportionately, but they frequently appear to sample
different sections of the same population. This applies not only to the
'attractant'-based trapping methods, but also to many 'non-attractant'
sampling techniques, which can therefore no longer be regarded as completely
unbiased.

The object of the present book is to review existing knowledge about
these essential field study methods which are the basis of existing knowledge
about ecology and behaviour patterns. Particular attention will be paid to
those investigations in which critical experimental studies have been carried
out on the validity of the capture techniques employed, or in which several
different capture or sampling methods have been tested concurrently on the
same species or group of species. Both methods of approach have produced
new ideas of the greatest significance in vector studies in general.

The fact that many essentially similar problems have been tackled in
quite different ways by workers in separate specialised fields unaware or out
of contact with each other, also makes it essential from time to time for some
non-involved outside observer to attempt an overall assessment of common
aims and common threads. In attempting such an assessment of a subject
whose world-wide importance manifests itself in the hundreds of new publications each year, it is necessary to be very selective in order to keep the
text within bounds. In this process of selection, with its special emphasis
on advances in knowledge in relation to the basic study methods employed,
it is inevitable that some noteworthy reports may appear to have been overlooked, or not given due recognition. This is to be regretted; but the
alternative, of attempting to include all publication without selection, would,
on the other hand, result in the sort of condensed uncritical summary of the
literature which would stifle itself.

In this selection of material dealing with behaviour patterns and the
methods used in their study, the subject matter has been confined strictly to
field study methods. No attempt has been made to incorporate the many
significant advances in the development of laboratory studies on the analysis
of behaviour patterns, and their interpretation in terms of insect physiology.
Such work has been particularly striking in the case of mosquitoes (Jones and
co-workers, 1967; Jones and Gubbins, 1974) and tsetse flies (Huyton and
Brady, 1975), and ideally should be considered as an essential component of
any general discussion on behaviour patterns. But here again, the rapid
progress in such laboratory studies would make it extremely difficult to do
justice to that discipline within the scope of this book. This omission does

not imply in any way an underestimate of the relevance of this work to field conditions.

Already sophisticated and accurately controllable laboratory techniques have done much to clarify the underlying causes of such recurring phenomena as flight rhythms and biting cycles of mosquitoes and tsetse, and there is little doubt that a much closer accord between field and laboratory experience regarding the same basic aspects of behaviour will play an increasingly important role in the future.

CHAPTER 2

Mosquitoes in Rural Settlements

i.   Introduction

Rural settlements and villages in tropical countries can provide ideal conditions for the transmission of disease. A wide range of blood-sucking insects are attracted to man and his associated domestic animals, and many of these also utilize the shade and shelter provided by huts and stables. Domestic and animal offal provides an ideal breeding ground for house flies, stable flies, blow flies and others. In addition to the natural surface waters such as rivers, pools and ponds often associated with such settle-ments, human activities create a range of artificial water collections such as dams, rice-fields, borrow-pits, wells, ditches, etc., all of which provide a wide choice of breeding grounds for many different mosquitoes.

Rural houses, villages or settlements have long been recognized as the main foci of malaria transmission in many tropical and subtropical countries (Figures 2.1 and 2.2). When these villages grow in size, or are absorbed in the sprawling suburbs which form the fringe of so many rapidly expanding towns and townships in developing countries, essentially the same housing units may also provide attractive conditions for the mosquitoes which transmit urban filariasis, yellow fever and dengue, mosquitoes which all tend to breed prolifically in man-made collections of water.

Fig. 2.1   The environment of rural dwellings in Liberia, typical of large parts of Africa.

Fig. 2.2 Rural housing and environment in the Philippines.

ii.   Mosquito Vectors of Malaria

For many of the most important malaria-carrying mosquitoes, the human settlement not only provides a source of blood meal — human or animal — during the nightly invasion, but it also provides ideal daytime resting conditions for the mosquito after its blood meal, in the dark corners and crevices of the walls and thatch roof. This tendency to remain within the shelter of the house after the blood feed is characteristic of some of the most important malaria vector species, both in Africa (for example, Anopheles gambiae and An. funestus) and in Asia (An. minimus and An. culicifacies). In contrast to these 'domestic' anophelines, there are other equally important species such as An. maculatus in Malaya, An. balabacensis in the Far East, and An. albimanus and An. aquasalis in Central and South America, which enter houses and animal sheds to feed at night, but only remain there to a limited extent during the day. Between these two extremes there are various gradations which may indicate fundamental differences in behaviour pattern between species, or which may be a consequence of an unsuitable environment — such as that in a well-illuminated house with open doors and windows or fenestrated walls — failing to provide suitable dark shelters (Figure 2.3).

Because of the domestic habits of many of the principal malaria vector species in the Old World — where so much of the pioneer work on mosquitoes was carried out — the house and its environs has been the main study centre for the capture, trapping and observation of mosquitoes. With all its attractive features, the village focus forms a concentration site for mosquitoes from a wide radius around, and sometimes from distant breeding grounds as well. It is for this reason that village houses and associated animal shelters have become the main centres of attack against adult mosquitoes in malaria control operations, in which all indoor resting places and surfaces are treated with residual insecticides in such a way that they remain lethal to mosquitoes which make contact with them, treated surfaces remaining effective for weeks or even months after each treatment.

Mosquito studies in and around houses have two main objectives. The main concern may be to establish basic behaviour patterns and population densities under natural conditions untouched by insecticides. On the other hand the main emphasis may be on studying the impact of control measures on the mosquito population, and its capacity to maintain malaria transmission. In many countries, in which malaria control or eradication programmes have been carried out over many years, it is becoming more and more difficult to find completely uncontaminated rural settlements or villages. Even in the absence of adult mosquito measures, there is always the possibility that the area may have been exposed to insecticides from other sources, for example in the course of aerial spraying against agricultural pests, or — in parts of Africa — against tsetse flies.

Whether the objective is the establishing of natural behaviour patterns, or the study of mosquito movements and populations in relation to control measures, the various methods used by entomologists are essentially the same, and may be briefly summarized as follows.

Live-catch of daytime resting mosquitoes by means of a suction tube or aspirator. This is a long-established method, the advantage of which is that the live mosquitoes caught are in good condition for identification and can be kept alive in cages until they lay eggs or develop malaria parasites to the infective stage. In those cases where the members of 'species-complexes' can only be distinguished by cytological characters only observable at a certain stage in the development of the ovaries (see page 73) mosquitoes which are fully blood-fed at the time of collection have to be kept alive for the appropriate period of blood digestion and ovarian development to elapse. While the process of collecting individual mosquitoes by suction

Fig. 2.3 Rural dwellings in
El Salvador, Central America, showing
open 'fenestrated' walls.

tube may on occasions be a tedious one, nevertheless valuable information is obtained about the exact resting sites of mosquitoes by day, whether in association with human and animal shelters or in natural outdoor sites.

A great deal of the recorded literature about the natural infection rates of the principal malaria vector species with regard to stomach infections (oocysts) and salivary gland (sporozoite) infections, is based on dissections of mosquitoes caught by suction tube and on catches carried out over many years and covering all seasons.

When these indoor live catches of daytime resting mosquitoes were standardized in order to compare mosquito densities from house to house, or from month to month, such figures — in combination with the infection rates of the mosquitoes — gave the first indication about which season of the year could be regarded as the time of maximum malaria transmission. This period was characterised by high densities of house-caught mosquitoes together with a high proportion of female mosquitoes containing infective sporozoites in their salivary glands. At the other extreme, a combination of low house-catches and scarcity or absence of infections in the caught mosquitoes, indicated a period of negligible transmission marking the 'non-malarious season'.

From these simple study methods has gradually arisen the whole concept of the seasonal periodicity of malaria which now plays a vital role in the strategy of malaria control by measures directed against the adult mosquito.

Space spraying and knock-down catch. While the routine hand-collecting of mosquitoes resting indoors is a useful practical means of obtaining live samples of the population, two obvious limitations had long been recognized in its use as a measure of mosquito resting density. Firstly, many rural dwellings in tropical countries have high ceilings or roofs, with many dark corners not easily accessible to the mosquito collector. Secondly, hand-catching becomes increasingly time-consuming and inefficient at seasons of the year when mosquito production is at its lowest, and the average village house may only yield one or two mosquitoes in an hour's search.

In order to deal with such situations, the method used is to cover all windows, doors and other openings with tarpaulin sheets or other opaque material, and then spray the interior of the house with pyrethrum aerosol. This produces a rapid knockdown of all resting mosquitoes, even those in the highest angles of the roof, and these can then be picked up from white sheets previously spread over the entire floor, as well as over beds, tables and other horizontal surfaces.

The most striking disclosure of space spraying is that it reveals the limitations of hand-catching, even under ideal conditions, as a measure of resting densities. At best the hand catch may only detect 50% of the resting mosquitoes, and this figure may fall as low as 10% in houses with high roofs and many inaccessible corners. The discrepancy is most marked at low mosquito densities, when many houses which have been declared negative after hand-catching still produce significant numbers after space spraying.

The most useful application of these two techniques is to carry out a preliminary hand search in order to produce live mosquitoes for various purposes, and then follow this in the same house by space spraying in order to reach a more accurate figure for the total number of resting insects.

Window traps and exit traps. When samples of the house-resting mosquito population are examined, the females — which alone suck blood — can be arbitrarily classified into four groups: unfed, freshly engorged with blood, half-gravid and gravid. This is in accordance with the fact that with most tropical species of *Anopheles*, it usually takes 2-3 days for the blood meal to digest and for the ovaries to develop to full maturity. For examples, a female feeding on Monday night would be recorded as freshly-engorged when captured indoors early on Tuesday. By Wednesday morning it would have reached the half-gravid stage, and be fully gravid by Wednesday night. The gravid female may be sufficiently developed to stimulate it to leave the house that night and look for a suitable water collection in which to lay its eggs. But frequently oviposition may be postponed until the following night. In Thursday morning's daytime collection, those females would be classified as fully gravid.

After oviposition, the female returns for a further blood meal — possibly later the same night — and the cycle of blood digestion and development of ovaries, the gonotrophic cycle, starts all over again. The unfed fraction of the mosquito house catch may contain recently emerged females not yet ready to take a blood feed, as well as older hungry mosquitoes which for some reason or other were not successful in obtaining a blood meal.

The proportion of mosquitoes in these different stages of the gonotrophic cycle can provide the first useful guide to movements during this 2-3 day period. If the daytime collection regularly reveals that the numbers of half-gravids and the numbers of fully engorged remain roughly equal, the indications are that the female is remaining indoors for the greater part of the cycle. On the other hand, if the proportion of engorged females is consistently higher than that of the gravids and half-gravids, the conclusion is that many females are leaving the house before the end of the 2-3 day cycle.

Even under ideal resting conditions, many of the most 'domestic' anopheline mosquitoes reveal a consistently higher proportion of engorged as compared with gravids, showing that there is considerable egress in the earlier part of the cycle. Where indoor resting conditions are not ideal, the proportion of engorged females is so much in excess of the later stages as to indicate that the great majority of engorged mosquitoes leave the shelter of the house at daybreak. With many of the less 'domestic' anophelines, this is the normal pattern of behaviour; such mosquitoes showing very little tendency to remain indoors during the day, even when abundant dark sheltered resting places are available.

It should be noted that these conclusions regarding mosquito movements in relation to housing are based entirely on samples of resting mosquitoes, and they assume that the only mosquitoes entering houses are those seeking a blood meal on the occupants. Clearly, these assumptions would be invalid if there was a considerable degree of house entry by mosquitoes after they had fed elsewhere, outdoors or in some other habitation. House entry by females other than those seeking a blood meal can usually be confirmed or ruled out, as the case may be, by parallel collections in unoccupied houses, But here again negative results may not be entirely conclusive, as the temperature and humidity conditions in an empty hut may differ significantly from that of a habitation occupied by a whole family and their associated domestic animals.

The first logical follow-up of these relatively crude observations is to devise a method of trapping or sampling mosquitoes leaving the house during the gonotrophic cyclie. The basic principles of such window traps or exit traps is that mosquitoes leaving a house at dusk or dawn, as well as the hours of darkness in between, are strongly attracted to the light — even the faint light of the night sky — coming in through windows or other openings in the walls or eaves of the habitation. Simple lobster-pot types of mosquito netting trap placed over such openings prove very effective in attracting and retaining egressing mosquitoes (Figures 2.4, 2.5 and 2.6). Window traps of this kind can be installed in normal village houses, but there are obvious limitations due to the uncontrolled movement of the occupants in and out of doors. Also, their efficiency may be reduced by the existence of numerous other small gaps and fissures, especially between the walls and the roof, which provide alternative escape routes. Consequently, experimental mud and thatch huts of uniform size and construction, occupied by hired humans who act as mosquito lures or bait, are best suited for critical work in this direction. These huts are made in such a way that when the door is

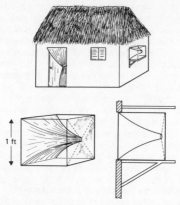

Fig. 2.4   Details of window trap

Fig. 2.5   Mud-walled, thatch-roofed experimental hut fitted with window or exit traps.

Fig. 2.6   Window trap fitted to occupied rural dwelling in Jamaica.

closed and light-proofed, the only light coming into the hut is through one – or two – small window openings provided with exit traps. Hungry mosquitoes enter the nut through the innumerable small openings in the eaves – openings too small to allow obvious light to penetrate from outside. Once inside the hut, the only obvious visible means of egress are the window openings over which the exit traps are placed. Even on the darkest night, these window openings provide a faint, but obvious, light attraction.

The basic design of experimental hut and window traps has undergone many modifications at different hands, but in one form or other it has been applied to most of the anopheline vectors of malaria throughout the world. The technique has yielded an immense amount of information about different behaviour patterns with regard to house entry and exit. In addition it has proved a valuable method of trapping or sampling many different species of anopheline mosquito, as well as culicines, which enter houses or animal shelters to feed on the occupants at night, but leave at or before daybreak.

Sampling the outdoor resting population. Those species of malaria vector mosquitoes which leave houses or animal shelters after feeding on the occupants, and which do not then enter other habitations, must spend the greater part of each gonotrophic cycle in natural outdoor resting sites, among vegetation, in shady banks and crevices, in outress roots of large trees, in animal burrows, etc. Among some such exophilic species, a good example is *Anopheles aquasalis*, main vector of malaria in Trinidad and coastal regions of Venezuela and Guiana, of which only 2% of the house-entering population remains there after daybreak. The collection and sampling of the remaining 98% has been based on painstaking searches outdoors in a variety of natural vegetation. As described above, even the notorious domestic, or endophilic, species such as *Anopheles gambiae* and *An. funestus* in Africa do not all remain indoors during the entire gonotrophic cycle, varying proportions leaving the shelter during that 2-3 day period. Some of those species show a preference for shady depressions or hollows, and for crevices in earth banks, and this observation has led to the development of artificial concentration sites or pit snelters (Figure 2.7) which have played an important part in clarifying outdoor behaviour patterns, and in providing comparable methods of sampling those populations in different areas and at different seasons of the year (Muirhead-Thomson, 1958, 1968).

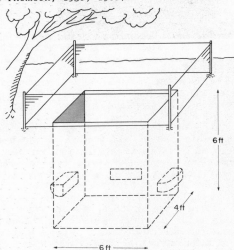

Fig. 2.7 Diagram of pit shelter showing dark cavities or niches

Study of biting populations.   One of the main characteristics determining the importance or otherwise of different species of anopheline mosquitoes as vectors of human malaria is the degree of association with man as a regular source of blood meal.   It follows that the collection or trapping of mosquitoes attracted to human bait is a widely practised method of sampling sucn mosquito populations.   When these methods are carried out in a standardized and comparable manner, valuable information can be obtained about behaviour patterns not only with regard to the relative association between man and different species of mosquito, but also with regard to man/ mosquito relationships at different seasons of the year, different periods of the day and night, and also the extent to which this association is affected by different mosquito control measures.   Groups of local village people may provide the attractive bait, or the mosquito collectors themselves may provide the attraction.   Parallel collections of mosquitoes attracted to human bait and to different kinds of domestic animal give information about which species are essentially anthropophilic, i.e. attracted to man, and which are zoophilic, showing preference for animal blood.

For many years such biting collections have played a prominent part in mosquito investigations, providing a wealth of information about this aspect of mosquito activity.   Particularly illuminating examples will be described in more detail in the course of this book.   In the context of this particular section about mosquito behaviour patterns in relation to human habitations, human bait collections provide valuable information supplementary to that obtained by the house collection of resting mosquitoes and window trap samples.   Biting activities at different periods of the night can be established both indoors and outside the house, and this information can then be related to what is already known about house entry and house exit of hungry and of engorged mosquitoes.

iii.   **Anopheline Mosquito Studies in the World Health Organization Malaria Eradication Programme**

The malaria control and eradication programme coordinated by the World Health Organization is based primarily on the treatment of all inside surfaces of human and associated animal shelters with residual insecticides (Fontaine, 1978).   Mosquitoes resting on these treated surfaces, whether walls, ceilings, under surfaces of the roof, etc., are liable to absorb a lethal dose of tne insecticide.   Depending on the nature of the chemical, and the material composing inner walls and ceilings (mud, brick, thatch, etc.), the deposit from one treatment may remain active for many weeks, or even months under favourable conditions.   These measures aim to reduce the mosquito population, and also to ensure that any survivors are unlikely to live long enough to develop malaria parasites, which they have ingested in the course of a blood feed on an infectious human, to the infective sporozoite stage in the salivary glands.

The vital importance of the house and associated shelters as the focal point of anopheline mosquito control has stimulated the need for more precise information about these mosquito movements in and out of inhabited dwellings, the length of the period they normally spend resting both before and after the blood feed, and the mortality among mosquitoes inside the house itself and among those which are able to leave treated houses.

One of the main techniques adopted by the WHO is the experimental mud and thatch hut with window trap attaches, as described above.   New insecticides or new formulations of possible value in the control programme pass through various laboratory screening tests.   Those that emerge successfully from the first three of these stages are then field-tested for the first time in a series of experimental huts constructed in areas where natural malaria vectors exist in densities sufficiently high to ensure high house entry rates for most of the year (Wright, 1971; Jurjevskis and Stiles, 1979). The main centre of these stage IV tests was for many years Arusha in Tanzania, where the malaria vectors are *Anopheles gambiae* and *An. funestus*. Another test ground for the same vector species is at Bobodioulasso, Upper Volta, in West Africa. A third centre in the central American republic of El Salvador has been mainly concerned with the important malaria vector of that region, *Anopheles albimanus*, a mosquito which has proved particularly resilient to a sustained control programme of over 10 years duration.

The main objective of the stage IV tests is firstly to find out how malaria vector species of mosquito react in nature to different insecticide treatments — as a progressive step beyond the earlier stage tests which use laboratory colonies of mosquitoes — and secondly to find out the fate and persistence of insecticide deposits on local building materials, which may vary considerably in absorptive capacity. Three types of experimental huts are normally used: (i) mud walls and grass roof; (ii) plywood walls and grass roof, and (iii) huts with roofs and walls lined entirely with sorptive mud. The kill of mosquitoes entering treated huts is made up of the daily count of dead mosquitoes on the floor, plus the number in the exit window trap which die within 24 hours.

The question of the normal pattern of mosquito entry and exit in unsprayed occupied huts or capture stations has been studied by a WHO team in Kaduna, Nigeria (Rishikesh and Rosen, 1976). In this case it was found advantageous to study entry and exit in separate huts. In the entry experiment mosquitoes resting indoors between sunset and sunrise were caught by hand every alternate 10 minutes, for 30 minutes each hour. In the exit studies, exit traps were installed in the door, and emptied every hour until sunrise. The general pattern of entry and exit is shown in Figure 2.8.

When these mosquito collections were classified according to unfeds, feds, and gravids, a more complex situation was revealed showing a continual movement in and out of the huts during the night of all blood-digestion stages. It appears that the simple original concept of entry and exit outlined on page 11 may be complicated by the fact that some of the entering mosquitoes, including newly emerged females unable to feed, as well as gravids and half-gravids, may be seeking indoor shelter from outside. It is also difficult to say whether engorged blood-fed mosquitoes taken in the early part of the night represent those which have managed to feed on the occupants prior to capture, or whether they also include some which have fed outdoors after sunset and are seeking an indoor shelter.

These and other investigations reveal the real difficulty in interpreting trapping and capture data in terms of behaviour patterns. In addition, small differences in the design of entry and exit traps, and in the individual execution of the experiments, may introduce other variables. For example, in the entry experiment, the presence of a collector with a torchlight switched on at regular intervals may deter some entering mosquitoes from feeding undisturbed on the occupants. In the exit experiments, the results depend on whether the exodus is confined to those mosquitoes which were actually resting in the hut before sundown — entry being blocked after that period — or whether a free ingress was allowed throughout the night, in which case the exit trap catch could include some mosquitoes which entered the hut and left after only a brief period.

With regard to the 'unfeds', it was long held that these represent mainly recently emerged females, not yet able to take a blood meal, but using the hut as a temporary shelter. However, other studies in Kaduna showed that of the unfed *Anopheles gambiae* and *An. funestus* taken in exit traps, a significant proportion were parous, i.e. they had previously undergone at least one gonotrophic cycle leading to oviposition, and that some of these mosquitoes were old enough to be capable of transmitting malaria (Self and Pant, 1968).

The vexed question of house entry and exit has been approached in a

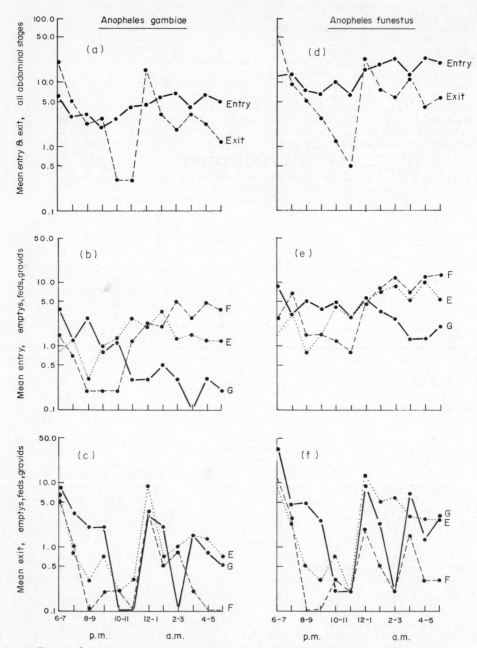

Fig. 2.8 Charts showing pattern of hut entry and exit by *Anopheles gambiae* and *An. funestus* in experimental huts with window (exit) traps. (After Rishikesh and Rosen, 1976).

different way in studies in El Salvador, with the main object of finding out how long the vector species, *Anopheles albimanus*, stays inside a house, an important factor in determining the success or otherwise of house-treatment with insecticide (Lassen and co-workers, 1972). This was done by: (i) charting cumulative curves of unfed mosquitoes entering and blood-fed mosquitoes leaving a house (Figure 2.9); (ii) marking mosquitoes, releasing them outside, and recording the number subsequently captured inside the house; (iii) marking mosquitoes after they had finished biting, and subsequently noting when they were captured in exit traps; and (iv) direct observation in ultraviolet light on mosquitoes coloured with fluorescent dye (Figure 2.10).

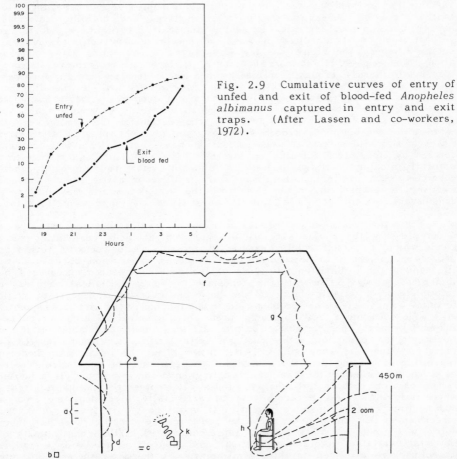

Fig. 2.9 Cumulative curves of entry of unfed and exit of blood-fed *Anopheles albimanus* captured in entry and exit traps. (After Lassen and co-workers, 1972).

Fig. 2.10 Results of direct observation in ultraviolet light of the behaviour of naturally entering unfed *A. albimanus* marked with fluorescent powder, 1970: (a) captured from outer surface of bobbinet curtain; (b) marked by dusting fluorescent powder, Lumogen BASF; (c) release of marked mosquitoes; (d) flight immediately after release; (e) a few minutes to $\frac{1}{2}$ hour later; (f) 1 to 2 hours later (some mosquitoes leave the house); (g) $\frac{1}{2}$ to 2 hours later; (h) 2 to 5 minutes biting; (i) rest for several hours; (j) bobbinet curtain encircling the house; (k) portable ultraviolet lamp and battery. (After Lassen and co-workers, 1972).

The results of these combined methods showed that over 80% of the *An. albimanus* that bit in the house stayed there for more than 5 hours. The direct observations made with ultraviolet light enabled movements of marked mosquitoes to be followed during flight and during resting periods inside the house, their movements in relation to approaching and feeding on the human bait, and any departures of marked mosquitoes from the hut during the night — the hut being entirely encircled by a bobbinet curtain.

These examples are only a few selected from the many studies carried out in recent years, the bulk of them concerned with evaluating the effect of house-treatment with insecticide on the behaviour and mortality of mosquitoes seeking a blood meal on the occupants. As described above, many of the variable factors encountered in a normal occupied rural dwelling (for example, size of house, nature of wall surfaces and building materials, number of occupants, facilities for entry and exit of mosquitoes, etc.), can be controlled or standardized by means of specially constructed experimental huts. However, there are obvious advantages in extending those observations, where possible, to normal occupied village houses (capture stations) occupied by family groups, especially when these families are fully cooperative in allowing access to collectors at different periods of the day and night, and allowing observations to be carried out on mosquitoes attracted to ot attempting to bite different members of the family. If window traps are inserted in such natural dwellings, the cooperation of the family is also essential in ensuring that door openings, which might provide alternative means of mosquito egress at peak periods such as sunset and dawn, are kept covered by light-proof curtains or screens. The additional information from naturally occupied village houses is also of vital importance in those countries, such as India, where man and his domestic cattle share the same dwelling, and where under the same roof the vector mosquito may feed to a considerable extent on those animals as well as on the humans (Shalaby, 1966).

For many years entomologists in malaria eradication programmes were under pressure to provide some useful entomological index (man-biting rate, etc.), which by itself could measure changes in vector density, differing degrees of contact with the human host, reduced expectation of mosquitoes' life, and so on. Unfortunately much of the genuine research encouraged by this pressure was offset by the tendency to apply statistical methods to incomplete or unreliable data, in the hope of achieving some sort of magic formula. Many of these problems in mosquito behaviour are consequently still wide open for a new approach, preferably in untouched areas uncontaminated by control programmes. Despite all the information about mosquito movements in connection with housing — either with or without insecticide pressure — there is surprisingly little known about the relationship between the biting or resting densities recorded in houses, and the population density as a whole. Whether the capture, sampling or observation point is a house, human bait or outdoor pit shelter, these all represent concentration sites where mosquitoes tend to congregate at artificially high densities from a surrounding area of unknown extent. It would be of great value to know, for example, that in a study area the house-resting densities, or the house-entry in a group of rural houses, represents 1/100, 1/1000 or 1/10,000 of the total available mosquito population, and also from how wide a radius mosquitoes are attracted to this focal point. Furthermore, there is already plenty of evidence to show that the relationship between indoor resting densities and outdoor resting densities may vary from place to place and from season to season. Much more accurate information on these points will be required before it is possible to elaborate on the present very tentative ideas about behaviour patterns.

Other investigations bearing on this same general problem will be discussed in later chapters.

Most of the research on mosquito movements in relation to houses and habitations carried out in connection with malaria control and eradication programmes has naturally been concerned with evaluating the effect of house treatment with residual insecticides. In the early days of application with DDT, the almost complete disappearance of resting populations of anophelines from inside treated houses was attributed solely to high mortality produced by this exceptionally lethal insecticide. The introduction of exit window traps, and the extension of the experimental hut technique, soon revealed that in the case of DDT at least, house treatment produced a drastic impact on the normal flight or behaviour pattern. In many cases the apparent 'disappearance' of mosquitoes from treated houses was not just a manifestation of high kill, but appeared to be due in part to the irritant or repellent properties of the insecticide or its particular formulation. These characteristics not only repelled some approaching mosquitoes from entry, but also had an irritant effect on mosquitoes attempting to settle on indoor treated surfaces either before or after they had taken a blood meal. The exit window trap revealed considerable egress on the part of those mosquitoes; and judging by their subsequent survival, many of those irritated mosquitoes had not absorbed a lethal dose of insecticide.

These original observations with regard to early formulations of DDT have proliferated into intensive studies in many malarious regions of the world. In most of that work, the experimental hut and window trap technique has played a vital part in studying the behaviour, or 'ethology', of mosquitoes in relation to indoor treated surfaces (Elliott and De Zulueta, 1975). Because of these known behavioural effects on mosquitoes, as well as other possible effects as yet undetected, there are now real difficulties in trying to establish what are 'normal' behaviour patterns in areas where there have been continuous spraying programmes over many years. If the houses or habitations used for establishing base line data have any history of insecticide treatment, or contain building material which has been exposed to insecticide, observations on house entry and exit may be difficult to interpret. But perhaps even more serious is the effect of such treatments on the mosquito population at large. Prolonged anti-mosquito spraying of houses may eliminate susceptible strains, leaving only the resistant ones. In addition, highly domestic strains, biological races or sibling species may disappear, leaving behind a population indistinguishable morphologically from the original but differing in essential characteristics which determine its behaviour pattern and ecological requirements.

iv.   Culicine Mosquitoes

Village houses and rural settlements in the tropics attract many culicine mosquitoes. As in the case of the anophelines, some of these could be classed as 'domestic' in that they not only feed indoors on man or his domestic animals, but also use the same shelter as daytime resting places. One of the most widespread of these culicines is the common *Culex pipiens fatigans*, the vector of urban filariasis, and a mosquito which thrives in the sprawling suburbs of fast-growing tropical towns like Colombo and Rangoon. In such places the mosquito breeds profusely in any collection of foul water such as pools in blocked drains, open privies, sullage water, etc. This mosquito has been studied by many of the techniques just described: hand capture, knockdown spray, window traps, etc. The World Health Organization in its surveillance studies falls back on the long-established practice of using hand catches indoors per man hour followed by pyrethrum aerosol knockdown spray; these figures are then combined to give a 'house index', i.e. number of female *Culex* per house (WHO, 1971a). One significant characteristic of this mosquito is that it has become resistant to many of the insecticides which have proved effective against anophelines, and consequently it is still possible to find high day-time resting densities in treated houses, catches of several hundred per house being not unusual.

A closely allied species, *Culex vishnui*, with similar domestic habits, has also been intensively studies, particularly in India, in its role as a vector of a viral disease of man, Japanese Equine Encephalitis, or JEE for short (Pant, 1979). In one investigation routine collections of mosquitoes were made not only in indoor resting sites, but also outdoors on various types of vegetation and growing crops (Reuben, 1971a). They showed that in the hot dry weather all indoor indices, both resting and biting, showed a fall in accordance with the drying up of many breeding places. However, the fall was not reflected in the outdoor resting densities, not — as one might surmise — because of increased exodus from houses, but simply due to the concentration of outdoor populations in the few sheltered vegetation sites available. With the onset of the dry season the fields of sugar cane, which had formed shelter for resting mosquitoes, dried out, and the mosquitoes now tended to concentrate on other plants, such as betel, which still provided shelter.

The existence of concentration sites outdoors, and the fact that these are liable to change with the season and the nature of the vegetation, is a fact which makes it difficult to express indoor and outdoor samples in terms of relative abundance, or even to compare outdoor densities at different seasons of the year.

One of the most intensive studies on the biology, population density and disease relationships of *Culex fatigans* was carried out in Rangoon from 1964 onwards by a World Health Organization research team and visiting scientists. The major conclusions of that international effort have already been assessed and reviewed elsewhere (Muirhead-Thomson, 1968; Service, 1976). More recently, studies on that insect have been intensified in New Delhi, mainly in connection with obtaining the necessary base-line data for a projected control operation using the sterile male technique. Of particular interest in the present context, however, are the studies initiated in Texas in 1970 on what is virtually the same mosquito, or an indistinguishable member of the same complex, long known in the Americas as *Culex pipiens quinquefasciatus*, a vector of urban Saint Louis Encephalitis (SLE) in temperate areas of the USA. What makes that investigation almost unique is that it is one of the first attempts to study vital questions of seasonal density, emergence rates, validity of sampling, etc., in a completely isolated population (Hayes, 1975). A site was selected where the only source of *Culex quinquefasciatus* was a large water trough, filled with water and provided with ideal larval food. Apart from the provision of a roof to protect the tank from adverse effects of sun and rain, the mosquito population — in particular the adult population — was exposed to natural conditions. Pre-study surveys showed that there were no naturally occurring populations of this mosquito in the experimental area, and this was confirmed by checks in the tank itself which remained negative for *Culex* egg rafts. When complete absence of *Culex* breeding had been established, egg rafts from a remote area were transferred to the tank daily for 10 days, and thus a continuous isolated focus of *Culex* was established. From then on, the various components of the *Culex* population — adult and immature — were regularly checked by means of regular daily counts of egg rafts, by weekly samples of larvae and pupae, by weekly counts of adults emerging during a 24-hour period by means of nets placed over the entire breeding tank, capture by aspirator (sucking tube) of adult *Culex* taken in specially constructed artificial sites, and by capture by suction tube of mosquitoes coming to bite bait in the form of pigeons (birds being the natural hosts of many species of *Culex*).

As the daily emergence of females (and males) into this isolated population was known, some idea of the efficiency of capture data could be obtained. It is significant that both with the outdoor resting population and the biting collection, the numbers taken were always low in relation to the numbers of adults previously emerged. The biting effort only collected 1/20 of the daily female emergence. When it is considered that the total adult population at any one time was probably considerably greater than the daily emergence, then it appears that only a very small proportion of the total population is being sampled by all the capture methods used. This discrepancy could be partly explained by high adult mortality, but unfortunately plans to study longevity had to be abandoned. Possibly a more likely explanation of the low availability of the recently emerged adult population is rapid dispersal from the study area. In this connection it is significant that reports from elsewhere have shown that overcrowding of larval *C.p. quinquefasciatus* — such as occurred from time to time in the main breeding tank — produces increased flight activity of subsequently emerging females. It seems very likely that some of these unresolved questions could have been illuminated by the use of the intensive marking-release-recapture which was used so effectively in studies on *Aedes* in Bangkok, and which are described below.

The main lesson from these Texas studies is the great advantage of working with a compeltely isolated mosquito population in which there is exact information about daily adult emergence and the daily addition to the existing population. It is a method which might well be applied to anopheline mosquitoes such as *Anopheles gambiae* in which dense breeding occurs in small sunlit rain-filled pools and puddles, of a type which can easily be created artificially. By careful selection of appropriate season and suitable isolated area where natural breeding places of this kind are absent, this method of approach would seem to have distinct possibilities.

v. The other notorious house-frequenting culicine mosquito is the classic vector of urban yellow fever and of dengue, *Aedes aegypti*, which breeds in collections of rain water in innumerable natural and artificial containers (discarded or broken bottles, used tyres, etc.) as well as in water storage jars and pots within the house itself. Other closely-related species of *Aedes* involved in disease transmission breed in the small collections of water in tree holes and plant axils, and these species make full use of such places when they occur in close proximity to houses and rural settlements.

In some of the early malaria control projects involving house-spraying with insecticide, it was found that *Aedes* was highly susceptible and easily controllable. However, hopes of its complete control or eradication have not been fulfilled, and today *Aedes aegypti* and its domestic and semi-domestic allies still constitute a major public health problem (Chow and co-workers, 1977; WHO, 197b, 1980). In 1960-62, for example, *Aedes aegypti* was responsible for an epidemic of yellow fever in Ethiopia which caused 30,000 deaths. More recently there was an outbreak of yellow fever in Ghana in 1977-78, and 1978 was also marked by epidemics in Colombia and Peru in the Americas, and in Gambia and Ghana in West Africa. The high fatality rate (100%) in the Gambia outbreak was a timely reminder of the explosive potential of *Aedes*-borne disease. *Aedes aegypti* was also the vector responsible for the epidemics of dengue haemorrhagic fever of Bangkok, Thailand, between 1958 and 1964.

Due to resistance to DDT, dieldrin and other organic insecticides, many *Aedes aegypti* eradication projects have been unsuccessful, and this has frequently led to re-invasion of formerly *Aedes*-free areas. In addition, eradication programmes have proved increasingly costly. At present, emphasis is on ultra-low-volume (ULV) applications of a particular organophosphorus insecticide, fenitrithion, and the use of focal control to anticipate outbreaks of viral disease.

Because of its long recognized domestic habits — feeding and breeding

in close association with man and human habitations — it has been found
convenient to record *Aedes aegypti* densities in terms of presence or absence of
larvae in domestic containers, rather than as adults.

The three best-known measures are: (i) the House Index, or the percentage of houses and their premises positive for larvae;   (ii) the Container
Index, or the percentage of water-holding containers positive for larvae, and
(iii) the Breteau Index, or number of positive containers per 100 houses
(WHO, 1971a).   The Breteau Index is the one most widely adopted by the
World Health Organization, and this is frequently supplemented by recording
landing/biting rates of adult *Aedes* females on human bait and expressing
them in man/hour terms, this additional measure being facilitated by the
fact that *Aedes aegypti* is almost entirely a day-time biter.

In order to standardize data about *Aedes* densities, and to facilitate
presentation of this information in map form, the WHO uses an arbitrary
'density index' ranging from 1 to 9, whose relation to the three larval
indices is shown in Table 2.1.

Table 2.1

Density	House Index	Container Index	Breteau Index
1	1-3	1-2	1-4
2	4-7	3-5	5-9
3	8-17	6-9	10-19
4	18-28	10-14	20-34
⋮	⋮	⋮	⋮
8	60-76	32-40	100-199
9	>77	>41	>200

The occurrence of larvae in domestic and peridomestic water-holding
containers is undoubtedly a sensitive measure of the presence of *Aedes
aegypti*, and this information is supplemented in many cases by the use of
ovitraps, standard artificial containers which *Aedes* females find attractive
sites for egg laying.   But this still leaves the question of *Aedes* adult
density in a rather less satisfactory state.   In areas where there is high
larval density of *Aedes*, adult density as measured by the attack or biting
rate always seems disproportionately low (Nelson and co-workers, 1978).
This was exemplified in Durban, on the Natal coast of South Africa, where
dense breeding of both *Aedes aegypti* and *Aedes simpsoni* is of great concern
to the Public Health authorities because of the possible recurrence of the
dengue epidemics of former years.   In one particular dump for discarded
tractor tyres, in which collections of rain water produced enormous larval
and pupal densities, the attack/biting rate rarely exceeded 10-12 per man/
hour.   From this alone it would appear highly likely that zero adult density
could be recorded at a time when high pupal densities indicated continuous
productivity from breeding sites.

This descrepancy was well brought out in an *Aedes* control campaign in
Thailand (Pant and co-workers, 1974).   Before insecticide treatment started,
the average number of adults per man hour was 25.2.   During the first month
after treatment this was reduced to the low figure of 0.1, i.e. a 99% reduction in estimated adult density.   But in that period, larval density as measured by the Breteau Index was never reduced by more than 90%, and the
infestation rate never at any time reached zero.

Because of the continuing public health challenge of *Aedes aegypti* and
its domestic and semi-domestic allies, work on those mosquitoes has been
intensified in the last few years, particularly by the World Health Organ-

ization which sent two research teams to Bangkok in Thailand, and to Dar-Es Salaam in Tanzania.   Of the many aspect of *Aedes* biology studied, five have been selected as being of special interest in the present context, as they involve sampling methods and their interpretation.

(i)   In Dar-Es-Salaam a more critical study has been made on the diel periodicity of biting on human bait, over 99% of which takes place between sunrise and sunset.   The results how a typical bimodal curve, with a trough around the midday period (Figure 2.11).   Where such bimodal curves have previously been reported, it has been suggested that these represent peaks of biting by two different sections of the adult female population, viz. the young nulliparous females which have not yet had their first blood meal, and the older females which have had at least one blood meal leading to ovarian development and egg-laying.   However, critical examination of the samples in Dar-es-Salaam, and parallel experiments with the semi-domestic *Aedes aegypti* in Uganda, showed conclusively that the shape of the bimodal curve was not determined by parous and nulliparous fractions (Corbet and Smith, 1974). This also applied to the closely-related peridomestic vector of yellow fever, *Aedes simpsoni* (McCrae, 1972).

The four other examples all make use of the marking-release-recapture technique for illuminating particular aspects of biology and behaviour, as well as providing estimates of population density: (ii) the determination of the duration of the gonotrophic cycle (Pant and Yasuno, 1973);   (iii) to analyse the differential domesticity of *Aedes aegypti* (Trpis and Hausermann, 1975);   (iv) to estimate population density on a mass marking basis (Reuben

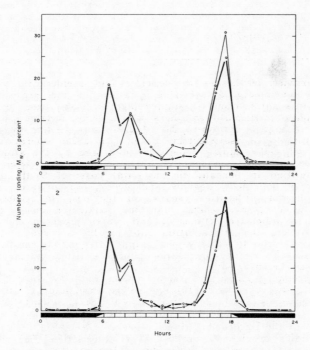

Fig. 2.11   *Aedes aegypti*: diel periodicities of landing on man.   1 — females
(thick line) and males (thin line);   2 — nullipars (thick line) and pars
(thin line).   (After Corbet and Smith, 1974).

and co-workers, 1973), and (v) to estimate populations and study flight patterns by means of individual marking (Sheppard and co-workers, 1969).

(ii) The cyclic pattern of blood feeding, development of ovaries leading to the gravid state, followed by egg laying, is essentially the same in *Aedes aegypti* as described for *Anopheles*. It is important to know the length of this cycle as it defines the normal interval between blood meals according to the season of the year. This frequency of feeding, taken in conjunction with the mosquitoes' expectation of life, determines the ability of the mosquito to acquire human pathogens or parasites in the course of the blood feed, and to live long enough for these to develop into infective stages, capable of being transmitted in the act of biting a new host.

In Bangkok, *Aedes aegypti* females of known physiological age (e.g. unfed virgin females and fed mated females) were marked with printers 'gold' and 'silver' fine powders and released indoors. The physiological condition of marked mosquitoes recorded in captures over the following few days indicated the normal period of gonotrophic cycle under those conditions.

(iii) Although *Aedes aegypti* in most parts of its distribution is highly domestic, there are parts of Africa where what appears to be typical *Aedes aegypti* is also recorded breeding remote from habitations, i.e. feral. This apparent anomaly was studied by collecting *Aedes* in three localities (domestic, semi-domestic and feral) and marking these with different fluorescent dyes (red for domestic, yellow for peri-domestic and green for feral) and then releasing them in a peridomestic habitat. Subsequent landing/biting collections made inside houses showed that, of the marked mosquitoes recovered, 83% were originally domestic, 15.5% were peri-domestic and only 1.5% feral. The conclusion from this is that there are three behaviouristically distinct populations of *Aedes aegypti*, in which the mosquitoes' characteristic of entering houses is a genetically controlled behaviour trait.

(iv) Mass marking, release and recapture of mosquitoes has been widely used in a variety of mosquito studies for many years, mainly with the object of determining flight range. In most of the investigations on anopheline mosquitoes recovery rates have been so small that valid conclusions have been difficult. In contrast to this, studies in a tyre dump area of intensive *Aedes* breeding in Delhi, in which batches of *Aedes* were marked and released on four successive days (a different colour for each day), produced a sufficiently high recovery rate to enable population estimates between 6000 and 12,000 to be made for the study area. In area terms this represents up to 22,000 *Aedes* per hectare.

(v) In Bangkok, marking, release and recapture has been carried out on an intensive scale by giving each marked mosquito an individual marking, distinct from all others. Individual marking of this kind has long been used in tsetse studies where the greater size and robustness of these insects allows several different colour marks to be made on the thorax. With the comparatively small and fragile mosquito body, this is a much more difficult operation, and consequently the successful application of this technique to *Aedes* represents a new and significant advance.

In an urban part of Bangkok, a section of 31 houses, each with several bedrooms, was selected. Indoor collections of adult female *Aedes* were made by suction tube or aspirator. Each mosquito was then given a unique mark in the form of a small spot of paint on one or other of eight selected sites on wings and thorax (Figure 2.12). One colour of paint was sufficient for a series of 255 consecutively numbered mosquitoes to be marked and released. Marked mosquitoes were released in the room of capture, and collections were made in rooms in the compound 3 times a week.

These studies provided a great deal of information about *Aedes* behaviour patterns, particularly with regard to direction of movement and distance covered between release and recapture. They also provided more precise data about the length of the gonotrophic cycle and the frequency of biting, and also with regard to the survival rate under local conditions. The mean

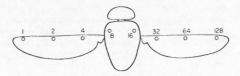

Fig. 2.12   Diagram showing the positions on the mosquito used for marking, together with the numerical
value of each position.   By using all the various combinations of positions, 255 consecutive numbers can be
given using a single colour of paint.   (After Sheppard
and co-workers, 1969).

daily population of *Aedes* females in the study area (94 x 56 m) was calculated to be 1093-1120, which in area terms is 5433 per hectare.
      Aedes aegypti has proved to be a particularly suitable mosquito for
such individual marking studies.   Movement is comparatively limited within
such an urban site, and consequently recovery rates are high enough to
enable valid conclusions to be reached.

      Mark-release-recapture experiments have also been carried out on *Aedes
aegypti* on the Kenya coast, based in a small village of 34 houses (McDonald,
1977a,b).   Adults were obtained from pupae collected in local domestic water
containers, and the mosquitoes were marked individually with a spot of
quick-drying synthetic paint.   With 12 different colours, which could be
applied to one side or the other of the thorax, mosquitoes were marked in 23
distinct ways for release in the houses in which they had been collected as
pupae.   Systematic landing/biting collections of mosquitoes were then carried
out in each of 20 houses in the village.   Marked mosquitoes recovered were
again colour-coded with the date of capture, and all mosquitoes released in
the same house from which they came.
      Of 119 marked captured females, 34 were captured twice, 5 captured
three times, and 3 four times.   With such a high proportion of re-captures,
survivorship curves for both males and females could be plotted.   The figures
in this first experiment showed that less than one-third of captured mosquitoes had markings, even though marking of newly-emerged females had
been carried out for 6 weeks.   There also appeared to be differences in
survival values of older mosquitoes and those mosquitoes in the first two
days post emergence.
      In order to test one possible explanation of this, namely that there
might be considerable movement between villages of newly emerged in particular, accounting for a lower capture rate, experiments were designed for the
simultaneous release of marked mosquitoes belonging to three different age
groups.   This was arranged by keeping emerging mosquitoes in the laboratory until they attained the appropriate age for release.   The houses in
which releases took place were fitted with exit window traps in order to
sample those mosquitoes leaving the house without making contact with the
human bait collectors.   Of 120 females for each age group (0, 2 and 4 days)
the total recaptured was 138, the percentage recapture for each age group
being exactly the same (38%), showing that the newly emerged mosquitoes had
no markedly greater mortality than the older ones, and this finding applied
to males as well as females.   These experiments differ from conventional
mark-release-recapture tests in that successive groups of mosquitoes entering
the population were marked, rather than samples of the adult population at
large.
      In further experiments to determine the extent of dispersal of *Aedes
aegypti* within and between villages, mosquitoes of three age groups and

both sexes were released at the rate of 120 (40 of each age group) inside each of 6 houses in the village.   Recaptures were made again by landing/ biting rates in all 32 houses in the village, supplemented by exit window traps.   The results showed that dispersal of both males and females occurred primarily to houses within 20 m, a high percentage of the recaptures being made in the house of release.

In the study of dispersal between villages, newly emerged adults were marked with fluorescent dusts and released at distances of 200 m, 400 m and 800 m from the centre of the village, at four different points of the compass. The results showed that the number of mosquitoes recaptured declined sharply when the release point was beyond 200 m.   For example, of nearly 1800 females released at 200 m in four experiments, 39–45% were recovered in the village, but this proportion fell to 2–10% for the 400 m release, and to 0.4–0.9% for those released at 800 m.   The greatest number of mosquitoes entering the village from outside release points came in from the downwind direction, and least from the upwind direction.   The high recovery rates of marked mosquitoes in all those experiments indicated a highly indigenous village population of *Aedes aegypti*, with minimal dispersion outside or between villages.

CHAPTER 3

Flight Patterns and Flight Paths of Mosquitoes

   i) Introduction. The study of wild (feral) species.
        Forest mosquitoes. Dispersal of swamp-breeding
        species. Parallel flight studies in the Gambia
        (West Africa) and in Florida, USA.
  ii) The Florida studies. Assessment of sampling techniques.
        Attractant and non-attractant traps. Truck traps,
        light traps and suction traps. Environmental factors
        at trapping sites. Visual barriers to flight.
        Direct observation on mosquito reaction to suction
        traps.
 iii) The Gambian studies in W. Africa. Ramp traps and inter-
        ception traps. Search flight and approach flight.
        Pattern of flight response to live bait and to CO_2.
        Flight paths in relation to height.
  iv) Other work on mosquito flight paths. American studies
        on comparison of light trap, CO_2 and other traps.
        Suction traps as 'non-attractants'. Electrocuting
        grids. Mark-release-recapture of woodland *Aedes*.
   v) Flight paths and resting populations. Observations on
        woodland *Aedes* in England.

i. Introduction

The sampling and study methods discussed so far have been concerned with establishing behaviour patterns of domestic and peri-domestic mosquitoes under conditions where man and his associated domestic animals, together with their habitations and shelters, provide the main focal centre of attraction.

Away from human settlements and domestic environments, intensive studies have been carried out on forest mosquitoes, with particular reference to the biting cycles of a wide range of both anopheline and culicine mosquitoes, and to their biting activities at different vertical heights from ground level upwards to the canopy. Some of the classic investigations — particularly those which formed part of Jungle Yellow Fever studies in Central Africa and South America — were established over 30 years ago, and have been fully reviewed and discussed. Work on those lines still continues, and some of the more recent findings will be assessed in another section. The main feature of all those studies is the emphasis on the biting fraction of the mosquito population, and the use of human or animal (both mammal and bird) bait to attract hungry mosquitoes, which are then either collected manually (by suction tube) or mechanically (by suction traps).

In some of these investigations, the main object of the intensive mosquito capture at different levels in the forest has been to provide adequate pools of the different suspect species of mosquito, on which tests for virus recovery can be carried out in order to incriminate the natural vectors. In other investigations, the same initial stimulus has led in turn to more general studies on behaviour patterns of both culicine vectors of virus and anopheline vectors of malaria, and eventually embraced other species not definitely incriminated as disease carriers.

Many of the mosquitoes discussed in the previous chapter utilize various types of water collection, created by human activity in and around human settlements, as breeding places. For such mosquitoes the recurring cycles of feeding, development of ovaries, and egg laying may involve comparatively short flights between feeding and breeding ground. There are other species, however, which breed in extensive natural areas of swamp land, both freshwater and brackish, often some distance from human settlements. Some of those breeding areas are very extensive and, consequently, under favourable conditions can provide an enormous output of mosquitoes, far in excess of anything possible from limited domestic and peri-domestic water collections. For such species the regular flight between the feeding centre (the village or human settlement) and the breeding ground may involve a long distance.

The dispersal of those high populations of mosquitoes from extensive breeding areas, and the manner in which their flight is directed towards the human and animal hosts, has long been recognized as a vital problem in mosquito biology and control (Gillett, 1979). Consequently there is a great deal of information about the extent to which these behaviour patterns are affected by such environmental factors as wind direction, climate and weather, and local topography. Any attempt to do full justice to the contributions of the numerous workers involved, or to summarize the work done in so many countries, could only defeat its ends by confusing the reader. Accordingly, two outstanding long-term studies have been selected for special attention, other relevant work being inserted where appropriate.

Each of these long-term studies — one in Florida and one in the Gambia, West Africa — have been concerned with the same basic problems of mosquito flight paths and flight patterns. But because they have been pursued more or less independently in two widely different geographical areas, with an entirely different range of mosquito species, the study methods have developed along distinctly different lines. As the Florida studies have been going on continuously for the longer period of 15 years or so, they will be discussed first.

ii.   The Florida Studies

Right from the start of the Florida work (Bidlingmayer, 1967, 1971, 1974, 1975; Bidlingmayer and co-workers, 1974; Bidlingmayer and Hem, 1979) great emphasis has been laid on the need to understand the limitations of each sampling method used in mosquito studies in order to interpret sampling data correctly. Their guiding principles was that "all samples regardless of the technique employed contain biases because of the differing behavioural pattern and responses characteristic of each species". The mosquitoes' response may differ according to its nutritional or gonadal state, whether unfed, engorged or gravid, and these differing responses will influence the availability for capture by different sampling methods. This is particularly marked in the capture methods using human or animal bait which may attract a certain segment of the population in the hungry or unfed physiological state.

Based on this approach, the work was carried out in three quite different study areas of Florida covering a wide range of brackish water, freshwater swamp and floodwater area, and embraced over a dozen different kinds of mosquito including species of *Anopheles*, *Culex* and *Aedes*.

An important distinction was drawn between attractant and non-attractant types of capture and sampling methods. All collections based on the use of bait (e.g. landing/biting and bait traps), as well as those using light (CO_2, etc.), use an attractant. In contrast are the passive or non-attractant methods in which the mosquitoes in the course of normal flight are caught or intercepted mechanically by netting screens (e.g. Malaise traps), by flight traps, by suction traps or by sticky traps. In theory the absence of an attractant should make it possible to obtain a completely representative and unbiased sample of a mosquito population in flight at the point of interception. Much of the Florida work has been concerned with the design of experiments to put this to the test.

In a first series of experiments four different types of sampling were used (Figure 3.1): (i) a truck trap, in which a large screen net mounted on a motor vehicle is driven along a prescribed route at twilight; (ii) the mosquito landing rate on bait; (iii) a suction trap constructed of plywood and equipped with a fan that moves approximately 100 cubic metres per minute, the trap being kept stationary; and (iv) a New Jersey light trap, widely used in the United States as a standard mosquito sampling method (Bidlingmayer, 1967, 1971).

The truck trap and the suction trap were regarded as non-attractant. During each twilight period, when there is normally an outburst of mosquito activity, the truck trap sampled approximately 25,000 cubic metres while the stationary trap sampled about 9000 cubic metres of air.

The results of this investigation revealed very clearly not only the existence of well-defined differences between the different sampling methods, but also the efficiency or performance of one attractant capture method in particular (the light trap) was greatly influenced by topographical features and intensity of moonlight. The suction trap, representing a non-attractant sampling method with minimum bias, revealed a higher level of mosquito flight activity at full moon than at new moon, whereas light trap collections were smaller at full moon. In addition a light trap located in the forest captured more than light traps in the open.

The very variable results obtained with light traps according to the phases of the moon has been reported from many other sources, and is perhaps not unexpected in that, while the light trap may emit a constant degree of illumination, this light intensity relative to the background light will naturally be at its lowest when competing with the bright illumination over the full moon period. In the same way the same intensity from the light trap will appear relatively brighter against the dark background of woodland or forest than in the open.

The differences between the effects of light (attractant) versus suction trap (non-attractant) is well brought out when different species of mosquito samples are arbitraily classified according to the ratio "New Jersey light trap : suction trap catch" during the entire night. The first group with a ratio ranging from 8:1 to 1:1 are highly attracted to light, and are exemplified by *Anopheles crucians*. The second group, with ratios ranging from 0.7:1 to 0.4:1, are only moderately attracted to light, the suction trap being more efficient. Examples are the brackish-water pest mosquito *Aedes sollicitans* and the swamp and floodwater breeding *Anopheles quadrimaculatus* (females), the main vector of malaria when that was formerly prevalent in the southern US. A third group were either not attracted to light traps or were actually repelled, giving an NJ:suction trap ratio of 0.2:1 down to 0.1:1, the last figure indicating that on average the light trap catch was only one-tenth of the suction trap catch. Examples of this were a pest culicine, *Mansonia perturbans*, and *Anopheles quadrimaculatus* males.

Later in these Florida studies additional attractant and non-attractant

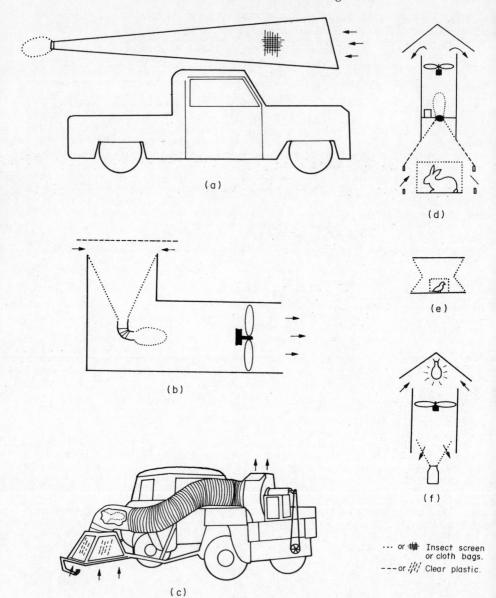

Fig. 3.1 Sampling equipment used in Florida mosquito studies. Arrows indicate direction of air flow. A – Truck trap. B – Suction trap. C – Vehicle aspirator. D – Baited suction trap. E – Lard can trap. F – New Jersey light trap. (After Bidlingmayer, 1974).

capture methods were used (Bidlingmayer, 1974). The non-attractant method took the form of a vehicle/aspirator (Figure 3.1C) which made collections of mosquitoes resting on ground litter as the vehicle moved over a fixed course. The attractant trap (Figure 3.1D), baited with hen or rabbit, was provided with a fan which operated in 'on' and 'off' periods in such a way that mosquitoes which had gathered round the bait were drawn upwards into a cloth mesh bag during the 'on' period.

Apart from the vehicle/aspirator, which was designed to sample daytime resting populations and only operated by day, all the other sampling procedures were carried out at fixed time periods throughout the night. This enabled the records to be related not only to unusually high periods of activity, such as dusk, but also to different phases of the moon.

Examination of female mosquitoes captured by these different methods according to their physiological condition (unfed, engorged, gravid, etc.) showed that both attractant-based methods (the bait trap and the light trap) were more selective than the non-attractant methods in obtaining represent-ative samples of the population (Bidlingmayer and co-workers, 1974). How-ever, even within the non-attractant group there is some evidence of bias in that the suction trap proved more selective than the vehicle trap, the latter taking not only the largest numbers of the flying population of most species, but also the largest proportion of engorged specimens. At the other extreme, the bait collection recorded the smallest number of blood-fed females.

Many mosquitoes have the highest level of flight activity during the twilight period, and the intensity of this crepuscular flight varies with different species. However, it was found difficult to obtain an exact measure of these differences because each sampling method provided a different ratio between crepuscular and dark period collections. For all species of mosquito, the catches from the truck trap operated on the highway adjacent to the stationary trap sites recorded higher crepuscular activity than the suction trap (non-attractant) as well as the bait trap (attractant). For some species, such as *Aedes taeniorhynchus* and *Aedes vexans*, the truck trap accounted for more than half the total catch taken throughout the entire night. In contrast, the New Jersey light trap took relatively low catches during the crepuscular period, with high catches in the latter half of the night and at dawn. The increased attraction to artificial light near dawn has also been reported by other observers working with different tropical mosquitoes, and is clearly an important factor in sampling studies.

Despite the many advantages of the vehicle trap, it suffered from one important limitation in that it could not be used in environments inaccessible by car. Consequently, the suction trap was chosen as the most suitable sampling method whose performance and efficiency could be further investig-ated in different types of capture site, such as in and adjacent to a wooded swamp and in the more open fields.

In the course of studies on these suction traps it soon emerged that, although they were identical in construction and operation, there were appreciable differences in the size and composition of the mosquito catches. This trap site phenomenon is well known from many other mosquito investig-ations; the information usually being put to immediate practical use by siting traps in the most productive situation. In few cases, however, has any deeper analysis of this site preference been attempted. The observations in Florida indicated that mosquito responses were influences by environmental factors at the different sites, particularly with regard to shrubs, trees and clearings. A series of experimenta were then designed to examine this, the L-shaped suction trap (Figure 3.1B) being used as the standard non-attrac-tant sampler (Bidlingmayer, 1975).

In the first experiment traps were arranged simply to determine the level of mosquito activity in the different habitats inside, adjacent to, and well outside a wooded swamp area. It had previously been shown that while the period of full moon was marked by greater mosquito flight activity than at the new moon — and for that reason might have seemed an ideal period for comparative tests — traps in the open were more strongly affected at that period than those beneath the forest canopy. Accordingly, for strictly comparable purposes, traps were operated for five nights only, centred on the new moon.

The next experiment, designed to test the effect of vegetation at the trap site, used shrubs about 8–11 feet tall planted in steel tubs so that they could be moved from site to site. Four traps were operated: firstly, without shrubs in order to test any bias associated with position of trap; secondly, with shrubs arranged in a close circle about 12 feet in diameter around the trap intake; thirdly, with the shrubs placed in a 30 ft row perpendicular to the edge of the swamp, the trap resting in the centre of the row as in a gap in a hedge.

The results of the initial check test (without shrubs) confirmed that mosquitoes could arbitrarily be divided into three classes: those which occurred in the greatest number in the open field, with numbers diminishing towards the interior of the swamp; those which occurred in the lowest numbers in the field, and in increasing numbers towards the interior of the swamp; and those with no well-defined habitat perference.

With regard to the experiments with shrubs, in nearly all cases the traps encircled with shrubs took smaller numbers of all species. In contrast, all species were taken in proportionately greater numbers in the traps placed in a row, or hedge, of shrubs, than when encircled. The reactions of different species are illustrated in Table 3.1.

Table 3.1   Ratio of suction trap collections in the field (= 1.00)
to suction traps associated with vegetation

	Swamp	Swamp edge	Field edge	Field	Hedge trap	Encircled trap
Psorophora ciliata	.08	.14	.50	1.00	.33	.14
Psorophora confinnis	.07	.16	.60	1.00	.48	.18
Mansonia titillans	.32	.45	.90	1.00	.54	.07
Anopheles crucians	.26	.38	.87	1.00	1.30	.24
Uranotaenia lowii	.67	.56	1.39	1.00	.86	.48
Uranotaenia sapphirina	.75	.70	1.55	1.00	1.14	.46
Anopheles quadrimaculatus	.77	.81	1.12	1.00	1.16	.53
Culex nigripalpus	3.36	1.92	1.67	1.00	1.66	.32
Culiseta melanura	5.31	3.23	1.38	1.00	2.51	1.11

Allowing for the usual day-to-day variations in catch influenced by climate and other environmental factors, mosquito reactions in these experiments were consistent enough to allow three arbitrary groups to be defined, this grouping being assisted by parallel collections of ground-resting mosquitoes carried out by day in representative areas.

In the first group, mosquitoes rest by day, mostly in open areas, with night flights occurring primarily in the field, The responses of this group to traps with shrubs indicate a response to physical objects independent of the illumination level. Whatever route was taken by mosquitoes in circumventing or ascending over the shrubs, avoidance of the shrubs took place at a sufficient distance to enable the field traps (unhampered) to take larger numbers of mosquitoes than either traps with shrubs.

In the second group, mosquitoes rest in shaded areas during the day. They fly more freely in the open area than in the swamp, the largest numbers occurring in the field edge traps along the outer face of the swamp.

Numbers taken by traps in the hedge were about equal to those of the field traps, but double those in the encircled traps. Those species will approach vegetation more closely than the previous group, but seem to resist penetrating foliage.

The third group rest in heavily-shaded woodland or swamps by day, and only relatively small numbers venture outside at night. The lower number taken in swamp edge traps than within the swamp itself shows that these species avoid the swamp margin and its higher level of illumination. With this group, hedge traps took larger numbers than field traps, indicating that these mosquitoes will approach shrubs closely and be guided into the trap intake.

Taking into account the general level of illumination at night according to the phases of the moon, and the effects of an overhead canopy of foliage, it was concluded that the general level of illumination is insufficient to limit mosquito movement, and that visual response to objects in the environment also determines habitat selection. These physical objects guide mosquito paths in three ways: firstly, by reducing the level of illumination, as beneath a forest canopy or adjacent to a forest edge; secondly, by forming a visual image to which each species responds according to its inherent behaviour pattern; and thirdly, by serving as a physical barrier.

As an extension of these visual barrier experiments, the tub-grown shrubs of the previous series were replaced by light-weight artificial barriers, 12 feet long by 6 feet high, so constructed as to fit around the suction trap roof. The frames of the barrier were covered by nylon fishnet: dark green cloth webbing, 2 ins wide, was woven in parallel strips through each row of mesh so that two-thirds of each square was obstructed but leaving the remaining openings sufficient to permit the passage of mosquitoes. When three of these net frames were stood on their long axes, with their bases one foot apart, they gave the appearance of being almost solid but in fact permitted free movement of mosquitoes through the mesh barrier. Three net frames were placed in line on opposite sides of the trap inlet, so that the trap fitted the gap between the ends of the six frames (Figure 3.2). These barrier traps were set up in the same range of sites in and near the wooded swamp as used in previous studies.

Mosquito response to an area of lowered light intensity was also investigated by using two large horizontal camouflage nets constructed so as to produce shade by means of green cloth webbing strands, while at the same time allowing sufficient openings for the free passage of mosquitoes (Figure 3.3).

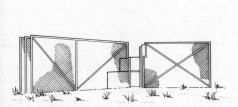 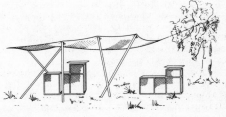

Fig. 3.2 Suction trap equipped with vertical visual barrier. (After Bidlingmayer, 1975).

Fig. 3.3 Suction traps, with horizontal net suspended above trap. (After Bidlingmayer, 1975).

Results showed that in the field sites nearly all species were taken in larger numbers when the traps were furnished with net frames.   In the wooded swamp, in contrast, the presence or absence of net frames had little effect on the mean number collected.   With regard to the overhead nets, this had its most marked effect on field species, whose numbers were sharply reduced.   Allowing for variables in response on the part of one or two of the several different mosquito species included in this test, these modified barrier tests showed that most species fell into approximately the same order of behaviour pattern as previously established.

The questions raised by the finding that large suction traps, theoretically non-attractive, may in fact be visually attractive to some mosquitoes have been further investigated by studying mosquito flight behaviour in the proximity of visually conspicuous objects (Bidlingmayer and Hem, 1979).

In the first of two experiments in Florida, two unpainted plywood suction traps that had weathered to a grey colour were used to test mosquito responses to immobile objects (Figure 3.4a).   The trap opening, 1.5 m above the ground and horizontal, measured 79 x 79 cm, each trap being furnished with a fan that moved 102 m^2 of air per minute.   In order to determine the effect of black traps which would provide a greater contrast with the background, one trap was covered with matt black plywood panels, except for roof, air intake and discharge.   Traps were operated in two different environments, one in a salt marsh area and the other about 10 km inland, in order to sample a wide range of mosquito species.

In addition to these two variations of the trap, a third trap with the same dimensions was designed with the object of determining the effect on catches by reducing the visibility of the trap.   This was done by constructing the trap of transparent acrylic plastic with the opaque material limited to the fan, fan mounting and aluminium screen used to concentrate the insects (Fig. 3.4b).

In the second experiment, two suction traps without roofs were buried 40 m apart in a dyke crossing an open salt marsh (Figure 3.4c).   Air was taken in at ground level and discharged at the side.   Four different modifications were then tested.   In the first the trap was provided with a rigid transparent plastic riser, 1.2 m high, with one end fitted into the trap opening.   Trap collections were made at an elevation of 1.2 m with the upper end of the riser serving as an air intake.   The second variation differed from the first in that the four outside surfaces of the transparent plastic riser were entirely wrapped in black cloth in order to make it clearly visible.   In the third variation, only the lower 0.9 m of the riser was wrapped in black cloth.   In the fourth variation, the riser was removed and the trap was furnished with a transparent baffle formed by joining two 1.2 m high sheets of acrylic plastic at right angles midway along the vertical surface, the air intake now being at ground level.   Within this design, traps were exposed either with transparent baffles, with one baffle completely covered with black cloth, or with no baffle at all.

In order to avoid any position bias in these comparisons of visible and 'invisible' traps, trap positions were changed each day, traps being operated from sunset to sunrise.

The results showed that in the first experiment most of the 15 different species of mosquito captured were taken in larger numbers in a weathered plywood trap covered with black cloth than in an uncovered trap.   When transparent traps were used, larger numbers of mosquitoes were captured in the panel-covered trap than in the transparent one.   From these results it appeared that black panels were attracting more mosquitoes.

In the case of the buried traps, results showed that responses varied according to species, but for each species the reactions were consistent.   For example, with *Culex nigripalpus*, a typical woodland species, traps with the completely wrapped riser and with the partly wrapped riser took 2-3 times as many mosquitoes as the transparent alternative.   Similarly, this species

Fig. 3.4 Suction traps in field location (experiment 1): (a) trap with detachable black panels is in foreground; subsequently both traps were located on a dyke in a salt marsh; (b) transparent acrylic plastic traps; the trap with detachable weathered plywood panels is in the background. The buried suction traps (experiment 2): (c) suction trap buried in a dyke; collections are made at ground level. (After Bidlingmayer and Hem, 1979).

preferred black baffles to either transparent baffles or no baffles.   In contrast, *Anopheles atropos* showed a consistent preference for transparent traps
over those with completely covered or partly covered risers.   Transparent
traps were also preferred by this species over those with covered baffles and
those with no baffled.   The reactions of other species were intermediate
between these two extremes.

The interpretation of these differences in reaction is that in the first
place visible objects are attractive to mosquitoes from a distance;   in close
proximity a change in flight direction occurs, with the adults of woodland
species approaching visible objects more closely than those of field species.
The different behaviour patterns close to the traps, with regard to attraction
or avoidance of different modifications, will also determine the extent to
which the close flight pattern exposes the mosquitoes to increasing chances of
capture by the effective suction zone of the fans.

Measurement made in still air of the air inflow 5 cm above and horizontal to the lip of the suction trap showed velocities of 347, 238, 145, 60
and 25 cm/sec at distances of 0, 10, 20, 30 and 40 cm respecitvely.   Laboratory experiments using flight chambers have provided data from several
different sources and with different species of *Aedes* and *Culex*, indicating
that the maximum flight speed ranges roughly from 150 to 250 cm/sec.   When
those figures are compared with trap-intake air velocities, it appears that
capture could be certain only within approximately 10 cm of the trap intake,
and decreasingly effective up to 20-30 cm and beyond.   Thus, if certain
species attracted to a visible suction trap usually pass within 30 cm of the
intake, a large proportion should be captured.   If other species usually
pass at a distance of 30 cm or more, fewer would be captured. Consequently,
visible suction traps can appear to be non-attractive, or even repellent, for
those species that seldom pass within the trap's effective radius.   Whether
suction trap collections of mosquitoes are increased or decreased by increasing a trap's visibility depends upon both trap design and the behaviour
response of different species.

The fact that the major differences in reaction of different species to
the range of traps provided are very consistent has provided sufficient
grounds for constructing diagrams representing presumed mosquito flight
behaviour with regard to four species in proximity to visible objects.   For
the moment these interpretations must remain somewhat speculative until such
time as they can be substantiated by direct observation on individual mosquitoes approaching these different traps.   While this may appear to be an
insuperable obstacle in the case of night-flying mosquitoes in the field,
technical advances may yet provide a solution.   Until that time concepts of
flight patterns must continue to rely on inference and on the correct interpretation of capture data.

The questions raised in this latest series of studies in Florida with
regard to the actual mechanisms of mosquito capture by suction trap are
very timely in view of the fact that these traps are playing an increasing
role in capture and sampling of many winged insect vectors of disease.   The
suction fan is also an essential component of such widely-used light traps as
the New Jersey trap and the CDC (Communicable Disease Centre) miniature
light trap.   This would perhaps therefore be a fitting place to interject some
highly significant studies on the same basic problem, also carried out in the
United States, but quite unconnected with the Florida work.   These two light
trap designs and other similar types long used in mosquito sampling all
embody a motor-driven rotary fan to draw attracted insects down into a
holding container suspended beneath the trap.   One of the long-recognized
disadvantages of these designs is that beetles and other heavy-bodies insects
are also drawn into the trap, where they damage the more fragile mosquitoes.
Provision of wide-screen mesh has not proved a satisfactory means of exclud-
ing such undesirables.   In order to deal with this problem, tests were
carried out in which the direction of air flow was reversed in such a way

that attracted insects are lifted *upwards* into a container above the trap (Wilton and Fay, 1972). Experiments with insectary colonies of *Anopheles albimanus* and *A. stephensi* were designed to compare, under identical conditions, the efficacy of this modified up-draft form with the conventional down-draft form (Figure 3.5). A range of air flows was used (171, 137 and 110 m/min).

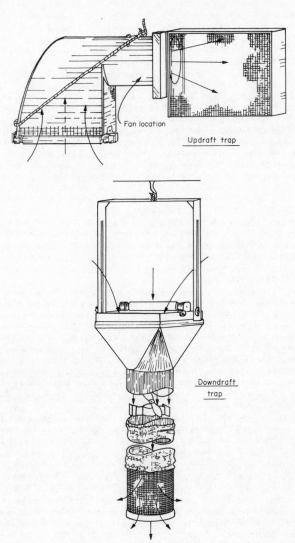

Fan location

Updraft trap

Downdraft trap

Fig. 3.5 Experimental traps used to compare the effects of upward and downward air flow on mosquito captures. (After Wilton and Fay, 1972).

The results showed a convincing superiority of the up-draft principle, mean captures for *A. albimanus* for example ranging from 42% to 72% depending on trap elevation, as compared with the mean down-draft captures which did not exceed 28% of exposed numbers captured. Over the range of air velocities tested, reduction from the maximum to the minimum showed that female catches actually increased with decreasing air flow until the air speed was reduced by half. In the down-draft trap, reduced air flow reduced the catch of both males and females. Of particular relevance to the vexed problems of mosquito behaviour patterns in relation to air movements in the vicinity of the fan or suction trap were experiments in which the flight of adult *A. albimanus*, dusted with fluorescent powder, was observed by means of UV light. Observations through the window of the rest chamber during the operation of a down-draft trap showed that individuals which approached the light, but managed to avoid being drawn into the trap, characteristically escaped by flying upwards. That this upward flight would appear to be the normal reaction of mosquitoes encountering an air stream is supported by the observation on the up-draft trap that the flight pattern was again upwards, and not downwards as might be expected if the reaction was solely against the air flow. In a trap of the conventional down-draft design, the air stream must overcome the lift factor in mosquito flight. If the trap produces an upward moving air stream, however, the mosquito flight reaction contributes not to its escape but to its capture.

iii.   **The Gambian Studies in W. Africa**

For the last 12 years or so — that is a period coinciding with much of the Florida work described — essentially similar problems on mosquito flight paths and patterns have been going on in the Gambia in West Africa, about 200 miles inland from the coast, on an entirely different range of mosquito species. Two of the main objectives of this work have been to study the long-range orientation of West African mosquitoes to man and other warm-blooded hosts, and to complement the considerable knowledge already available about the vertical distribution of mosquitoes in equatorial forest areas by means of similar studies in open savannah areas (Gillies, 1969; Gillies and Wilkes, 1970, 1972, 1974a,b, 1975; Gillies and co-workers, 1978; Snow, 1975, 1976; Snow and Boreham, 1973, 1978).

Extensive freshwater and brackish-water mosquito breeding grounds produced high populations of mosquitoes which then converged, sometimes from a considerable distance, on the human settlements and villages situated in the savannan/cultivated area inland from the river. In the early part of that investigation much thought was given to the design of a trap which could effectively intercept mosquitoes in their flight from breeding ground to feeding ground. The trap had to be non-attractant, and at the same time be designed in such a way as to provide the minimum physical barrier to mosquitoes, allow them easy access into the trap, and effectively retain them there. The trap finally designed, known as the ramp trap (Figure 3.6), allows mosquitoes to enter a wide opening (6 x 6 ft), inside which their flight is guided up a sloping mesh-netted ramp into a cage fitted over the top of the ramp and provided with a horizontal entry slit which could be opened or closed. Mosquitoes trapped in this cage in the course of their night-time flight could be removed from inside the top cage by means of an aspirator or sucking tube operated through sleeves in tne side (Gillies, 1969; Gillies and Wilkes, 1972).

The flight of mosquitoes from the breeding ground in search of a blood meal can usefully be divided into wandering or search flight which brings them into contact with odours carried downwind from the host, and directed or approach flight which then leads them to the actual host or source of blood meal.

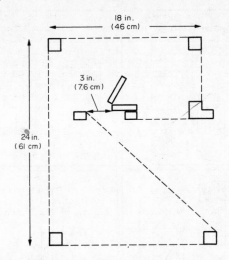

Fig. 3.6   Cross-section of the cage of the modified ramp-trap.
(After Gillies and Wilkes, 1972).

A series of experiments was designed to study aerial density of mosquitoes and the direction of flight in relation to wind direction and nocturnal
periods. Particular attention was paid to the problem of determining the
range over which natural or simulated baits affect the flight of hungry mosquitoes. In another series of experiments, units of three ramp or interception traps stacked one on top of the other to reach a height of 14 ft
above the ground were used to study the vertical distribution of the flight
paths of mosquitoes flying at comparatively low levels (Figure 3.7).   In a
third series of experiments, a critical comparison was made between ramp
traps and suction traps under identical conditions (Gillies and Wilkes, 1970,
1972;   Snow, 1975).
      In the centre of a cleared area near the mangrove swamp belt, which
extends far inland up the River Gambia, animal bait in the form of a single
calf or two calves was provided in a staked pen.   Radiating out from this,
in four directions to a distance of 50 yards, ramp traps were set up with
their openings facing away from the bait in such a way as to intercept mosquitoes flying towards it within 4-5 feet of the ground (Figure 3.8).   In
some experiments the animal bait was replaced by a CO_2 cylinder with
release control equivalent to the expelled breaths of the animal.   Catches
included a wide range of both anopheline and culicine mosquitoes.
      By plotting the density of mosquitoes in relation to distance from bait
in the form of a 'catch curve' it is possible to define a critical distance
at which mosquitoes start to converge from their wandering flight path.   The
results showed roughly three different patterns of response.
      The first group, which includes species of *Anopheles*, *Mansonia* and
Aedes, begin to show a response to a 2-animal bait at a distance of 20-30 m,
with a sharp increase at about 7.5 m.   With a single calf bait the range of
attraction is 9-18 m (Figure 3.9).   In the second category, dominated by
Culex thalassius and *C. antennatus*, the response is much the same but
begins at a much closer radius of 15 to 22.5 m of the host (Figure 3.10).
The third category is composed predominantly of bird-biting species of *Culex*
in which the trap density scarcely departs from zero with either bait.

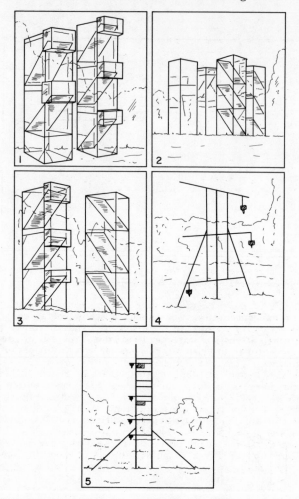

Fig. 3.7 Arrangement of flight (ramp or interceptor) traps and suction traps used in the Gambia.

(1) Flight traps at five levels (series 1).   (2) Four columns of three flight traps each (series 2).   (3) Two columns of three flight traps each (series 3).   (4) Non-directional suction traps at three levels (series 4). (5) The 9–15–m–high scaffolding tower, with suction traps at four levels (series 5).

(After Snow, 1975).

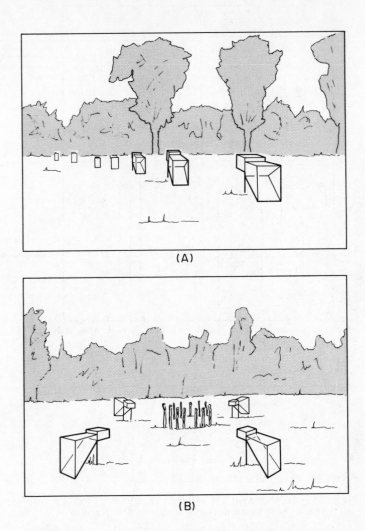

(A)

(B)

Fig. 3.8 Arrangement of ramp traps in test area (A) and control area (B) in relation to central animal enclosure.
(After Gillies and Wilkes, 1970).

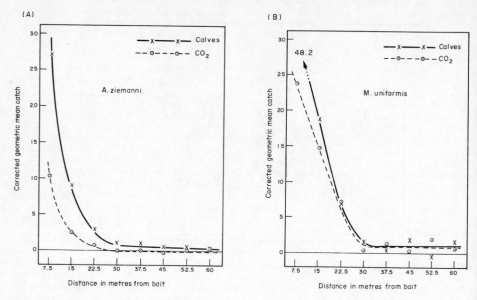

Fig. 3.9 Catch-curves for *A. ziemanni* (A) and *M. uniformis* (B) in
relation to distance from bait. (After Gillies and Wilkes, 1972).

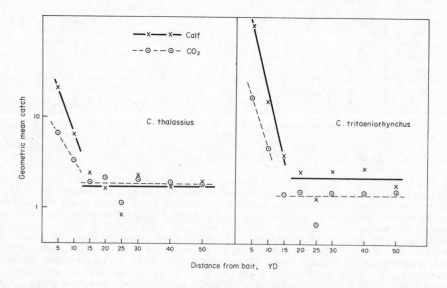

Fig. 3.10 Catch-curves of *C. thalassius* and *C. tritaeniorhynchus* in
relation to distance from the bait. (After Gillies and Wilkes, 1970).

This experiment with radiating traps also revealed consistent differences in flight direction, sometimes between closely related species. When the catch taken in the line of traps radiating out to the north and east were compared to those taken in the south and west, *Mansonia africana* showed a higher proportion in the north east trap (15.6) than any of the other species, including *M. uniformis* (7.2).

Field studies in the Gambia were then extended to a critical examination of the flight paths of different mosquitoes in relation to height above the ground. As the two main sampling methods used were the ramp or interception trap and the suction trap, this series of experiments also provided a valuable comparison of results obtained by the two different non-attractant techniques under identical conditions (Snow, 1975, 1976).

Flight traps were set over open ground at three levels (from 1.37 to 4.7 m) and the suction traps at either three or four levels (between 0.68 amd 9.15 m) (see Figure 3.7). A limited series of 4-level human-bait catches was also carried out at the same levels as the suction traps. With regard to flight levels, it was found that mosquitoes fell into three categories: (i) those most common at the lowest levels, such as *Anopheles*, *Mansonia* and the *Culex decens* group; (ii) those frequent at all levels, such as *Culex thalassius*; and (iii) those most common in the highest traps, mainly two ornithophilic species, *Culex neavei* and *C. weschii*.

Of particular interest was the disclosure in this series that under certain conditions the non-attractant ramp interceptor trap did have a visual effect on mosquitoes. For example, in the flight of *Anopheles melas* towards the bait in the centre of radiating arms of ramp traps there was no evidence of any trap avoidance, but when bait was absent parallel runs with suction traps showed that *An. melas* was avoiding the ramp traps. In contrast, the unbaited flight trap by itself was positively attractive to other species such as *Culex decens* and *C. thalassius*, particularly at moonlight, indicating that visual responses were involved.

These reports from the Gambia, taken in conjunction with the Florida experiences, highlight many accepted ideas about 'attractant' and 'non-attractant' techniques used in studying mosquito behaviour patterns. The absence of such obvious attractants as animal bait, CO_2 and light does not necessarily mean the elimination of other less obvious factors which can influence mosquito response. The framework of a trap, or the structure on which it is fixed, may be such as to present a visual object to which some species may be attracted. The presence of a trap may also have other physical effects; in the Gambia investigations, for example, it was found that the ramp traps or flight traps do not in general affect the wind direction, but that the wind speed is reduced by 50% immediately downwind from the trap.

iv.   Other Work on Mosquito Flight Paths

These general problems of mosquito flight and the interpretation of capture data have attracted the attention of many other investigators. Some of the results obtained are of particular relevance to the present discussion; for example, comparing different sampling methods simultaneously in order to reveal the advantages or the shortcomings of each method, as repeatedly emphasized by the Florida group. A great deal of work has been done in the United States comparing the effects of different light traps such as the New Jersey trap — widely used in the US, the CDC (Communicable Disease Centre) light trap, ultraviolet light traps, traps using CO_2 or dry ice as attractants, as well as light traps with and without supplementary CO_2 (Magnarelli, 1975; Vavra and co-workers, 1974; Wilton, 1975).

In one of those studies in Iowa (Acuff, 1974) two different light traps were compared with collections taken in an unbaited interceptor type of trap (Malaise), and also with collections attempting to feed on human bait.

In general the New Jersey light trap and the CDC miniature light trap baited with dry ice provided the best representative samples of the mosquito population — about 15 species of *Aedes*, *Anopheles* and *Culex*. One or two exceptional species were taken in the CDC trap but not in the New Jersey trap. The unbaited Malaise trap proved to be very selective for *Aedes* and *Culex*. The landing/biting collection was also selective for some species, while others were not represented. In the case of *Aedes vexans*, for example, 37% were taken by NJ trap, 61% in the CDC trap, 1% in the Malaise trap and 0.7% on human bait.

The miniature CDC trap supplemented by dry ice CO_2 has become widely accepted in the United States as the most efficient mosquito catching technique available. In terms of normal mosquito flight paths and behaviour patterns, however, it is very difficult to interpret light-trap data, as such traps introduce a new and disrupting factor into the environment. Nor do light traps provide a constant unvarying source of attraction in that there are great variations in the intensity of light, relative to the background, at different phases of the moon, and also according to the location of the trap.

Light traps have long been used to sample insects of agricultural importance in flight, and it is clear that those investigators too are faced with difficult problems of interpretation, rendered even more difficult in their case by the absence of comparable data from other sampling methods such as suction traps and truck traps, etc. As far as insects of agricultural importance are concerned, it has been concluded that "light traps thus obscure rather than reveal changes in activity and abundance because of the variation in trap effect with changing periods and amounts of moonlight" (Bowden and Morris, 1975).

The findings in the Gambia with regard to the importance of height above the ground at which sampling in savannah areas is carried out by flight trap and suction trap, have been endorsed in independent investigations (Service, 1971a). In one case it was found that the majority of mosquitoes in a particular study area flew close to the ground vegetation, and consequently suction traps with inlets 23–48 cm from the ground caught more adult mosquitoes than those placed at a greater height. That same study also provided additional information about the effects of visual objects in the surrounds and approaches of non-attractant suction traps. Traps with vertical strips of netting radiating out in three directions caught more mosquitoes than control traps, and it appeared that the strips of netting provided a visual guide to the mosquitoes, causing them to converge on the trap.

The visual stimulus provided by such 'non-attractant' flight or suction traps themselves, plus the additional visual attraction or repellence of the immediate surrounds of the traps, are obviously factors of considerable importance in determining how effective these techniques are in establishing accurate knowledge about normal mosquito behaviour patterns. The additional physical effect of large traps in providing a localised wind barrier and air eddies has also been a matter of concern in the Gambia work.

In order to minimise this wind-brake effect, electrified screens or electrocuting grids, which have been used so successfully in the tsetse studies discussed later, have been tested in the Gambia (Gillies and co-workers, 1978). The object was to use those grids as directional screens in association with suction traps, but certain limitations soon became apparent. Although the screens themselves have the advantage in that they do not produce any wind-drag effect, structures on which the traps and screens were originally fixed to operate at two levels above the ground (1 m and 2 m) did produce a wind-brake effect, giving anomalous results. Consequently all later tests were carried out at ground level, with their mouths just above the ground, and with supporting structures sunk in pits below ground level.

Another complication was that although the space between the wires of the grid was reduced to 5 mm in accordance with the smaller size of the mosquito as compared with the tsetse fly, some mosquitoes managed to pass through, and the grids were only 75-80% efficient in blocking the passage of mosquitoes.

Allowing for these limitations, it appears that suction traps in conjunction with electric grids can provide information regarding the direction of flight of mosquitoes upwind and downwind, and that it might be possible to dispense with the bulky suction trap component and use the screen alone as a trapping device, as has been established in the case of tsetse flies.

iv.   Other Work on Mosquito Flight Paths

The technique of mark, release and recapture has been extensively used to determine the flight range, distribution and longevity of natural populations of mosquitoes. In such experiments in the circumscribed environment of villages and rural settlements, the recovery rate of domestic species has in many cases been sufficiently high to draw valid conclusions about flight patterns. But this has rarely been the case with feral species released in woodland or forest habitats. An illuminating exception to this is provided by the following example.

Among recent studies involving mark-release-recapture of *Aedes* are those investigations of a species, *Aedes triseriatus*, which for many years attracted little interest from medical entomologists. This species breeds in tree holes and containers near the edge of deciduous forests (Sisinko and Craig, 1979; Scholl and co-workers, 1979). The adults, which seldom occur in large numbers, are diurnal and hence rarely taken in light trap collections. However, its incrimination a few years ago as the main vector of California encephalitis virus, and later of La Crosse virus, in mid-western USA, spotlighted the need for more information about this mosquito, particularly with regard to the size of natural populations (Sisinko and Craig, 1979). The workers involved reviewed all available knowledge about mark-release-recapture techniques for estimating mosquito populations, and came to the conclusion that all those studies were exposed to the same defect or shortcoming in that estimates of population size were made without any independent determination of accuracy. Any such assessment would require previous knowledge of the numbers of mosquitoes going into the population. In most cases this is not feasible under natural conditions. However, the particular ecology of *Aedes triseriatus* makes it possible to provide such an additional independent assessment. This is because discrete populations of this species can be found, and every pupa counted, in the known number of breeding places (tree-holes) within an isolated patch of woodland or 'woodlot'. The total number of individuals which have entered the population can be based on counts of total pupae found at frequent intervals. This total can then be used as an independent check on the estimate arrived at by mark-release-recapture techniques applied to adults. In obtaining pupal counts, the contents of each tree hole were completely extracted by means of a large pipette, and the pupae retained in the laboratory to indicate exact day-to-day emergence.

In order to collect sufficient adults for marking, various conventional methods were tried without success. In particular animal-baited traps, CDC light traps both with and without CO_2, and shiny traps all proved unproductive. The method finally adopted, utilizing the fact that *Aedes triseriatus* was biting man at a relatively high rate in that locality, was to use human volunteers and monitor the adult population by landing/biting collections.

Mosquitoes were marked by dusting with fluorescent pigments and identified by long-wave ultraviolet lamp. After the seven available colours of fluorescent dust had been used, mosquitoes subsequently captured were marked with coloured dots on the thorax.

Movement and dispersal between different woods was tested by marking the batches from different 'woodlots' with distinctive colours. Under these conditions the recapture rates for *Aedes triseriatus* were encouragingly high, at 29% of 1536 released, and there was no indication of mosquitoes leaving their particular woodlot. This provides an interesting contrast to another species included in these tests, namely *Aedes canadensis*, in which the recovery rate was only 2.7%, but in which 10% of recaptured mosquitoes had crossed from one woodlot to another. Recaptures of *Aedes triseriatus* within the 10 ha woodlot of release also showed that there was reasonably uniform mixing, with no tendency for this species to return to its capture site.

With regard to the particular woodland where mosquitoes were marked and released, mathematical models based on the unusually high recapture rates were used to calculate the total adult females in the woods at the end of the summer, and this gave a figure of 1225, indicating the comparatively small mosquito populations in this test area. The average daily survival rate was calculated from the same mathematical model, and compared with the figures of total females produced by pupae. The two independent assessments showed close correlation (Figure 3.11).

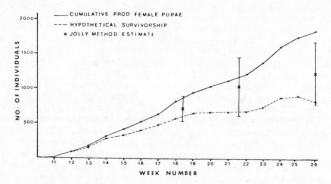

Fig. 3.11 Comparison of population estimates of *Aedes triseriatus* based on the Jolly method, and an independent check based on counts of female pupae. (After Sisino and Craig, 1979).

The results of these studies suggest that *Aedes triseriatus* populations tend to remain confined to their own particular woodlot, but move freely within the confines of that area. These workers point out that mathematical estimates of population size, as calculated by the Jolly method, were greatly assisted, and more fully substantiated, by the high recovery rates of marked mosquitoes — much higher than normally experienced in mosquito work of this kind. At the same time they point out that reliance on a sole capture method (human bait capture) has certain drawbacks. As biting activity was depressed by unfavourable climatic conditions, captures at such periods sampled smaller proportions of the existing population and consequently produced wide variations in population estimates.

v.   **Woodland Mosquitoes in England**

For a period of several years, coinciding with years of great activity in Florida, intensive investigations were being carried out in England on a group of mosquitoes which had previously received only scant attention, namely, the woodland species of both culicines and anophelines (Service, 1971a,b,c). One of these, *Aedes cantans*, is dominant in many of these patches of woodland, especially those in which suitable breeding habitats exist in the form of woodland pools. Humans entering such woodlands are readily attacked by these mosquitoes.

Studies on flight periodicity and vertical distribution were carried out mainly by the use of suction traps. Aerial populations were sampled initially by Johnson-Taylor suction traps and latterly by Vent-Axia traps, sunk into the ground so that the fan openings were at a heigh of 30 cm (Figure 3.12). These operated over 24 hours and automatically segregated the catches into hourly collections. These ground level catches were supplemented by suction traps operated initially at heights of 23, 73, 123, 173 and 223 cm, and latterly up to 550 cm. The traps were arranged so that expelled air from the fan was delivered to the side of the trap, and not directly downwards where it might have disturbed the air intake of the trap below (see Figure 3.12). The smooth intake and operation of fans was checked by smoke tests.

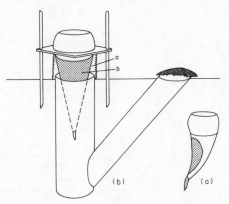

Fig. 3.12 Diagram of suction traps. (A) Fan and plastic cone of traps used to sample vertical distributions in 1968. (B) Suction trap placed in cylindrical tubing and lowered into ground for sampling vertical distributions in 1969 and 1979. (a, flexible tubing; b, fine wire mesh supporting the cone).
(After Service, 1971).

In all species, including *Aedes cantans*, unfed females constituted more than 96% of the total catch, with males normally 1% or less. Blood fed and gravid females made up less than 2% of the catch. These suction trap catches were supplemented by ten human bait catches from 16.00 to 22.00 hours. These two independent sets of data enabled flight periodicity to be compared with the biting cycle, and showed that with *Aedes cantans* the times of arrival of hungry unfed females at bait were similar to the flight times of unfed females as shown by the suction traps, with a peak from 19.00 to 20.00 hours (Figure 3.13).

With regard to vertical distribution, the suction trap experiment showed that females of *Aedes cantans* — as well as several other species investigated — which feed mainly on mammals, fly very close to the ground at 30 cm

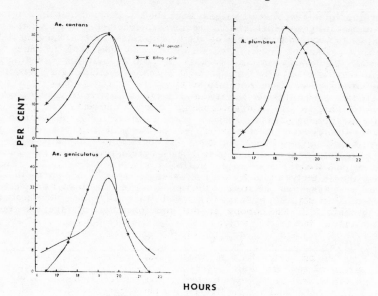

HOURS

Fig. 3.13   Flight periodicities and biting cycles of *Aedes cantans*,
A. plumbeus and *A. geniculatus* from 16.00 to 22.00 hours.
(After   Service, 1971).

or less, with less than 3% recorded at heights in excess of 100 cm.

Of particular interest was the extension of this study to include outdoor resting populations as an additional means of determining flight and distribution of mosquitoes. This was tackled in two ways: firstly, by the use of human bait captures carried out in a range of vegetation types, in order to study the movements of hungry unfed females; and secondly, by the technique of sweeping vegetation in order to determine the distribution of adults in all stages of the gonotrophic cycle.   The bait catches first of all showed that far fewer *Aedes cantans* were taken in the exposed sites selected than in the shaded sites.   It appeared that during the day, for this species at   least, unfed females rest in heavily shaded vegetation and are readily sampled in such sites by bait captures.   Further studies showed that within the general concept of favourable sheltered sites there was a further patchy distribution according to specially favoured loci, and that there was no random distribution of the unfed females.   Stationary bait in these sites initially attracted numbers of biting females, but these numbers declined until the bait moved into a new resting locus.

By the systematic use of transects, it appeared that unfed females resting in vegetation by day are attracted to a stationary bait over a distance of about 7 m, but not 10 m.   Furthermore, in all transects, the distribution of *Aedes cantans* conformed to a definite pattern which revealed that some of the sites along the transect are consistently more attractive resting sites than others.   Further experiments, in which morning bait captures were followed by afternoon ones in the same locus, showed that within these select habitats there is very little movement of unfed *Aedes cantans* during the day.

With regard to the technique of sweeping vegetation with a net by day, intensive searches were carried out over a period of 6 years.   In many of the vegetation types catches were of a low order, at best under 5 females per 100 sweeps. However, in a particular copse this figure increased to over 86 per 100 sweeps. These sweep net samples revealed females in all stages of the gonotrophic cycle, including blood fed (22.8%), half gravid (19.1%), and gravid (18.6%), the remainder being unfed.

By means of sweep collections carried out within a 30 m radius of cattle on which *Aedes cantans* had been seen biting, as well as by direct observation of gorged flies leaving bait, there is every indication that this species at least does not normally rest in the vicinity of the host on which it has gorged. Adults have been observed to fly off into the distance, and this distance may be considerable before suitable resting sites are found.

The existence of highly preferred loci within a general area of attractive and suitable vegetation may well help to explain why not only *Aedes cantans*, but also many other daytime resting mosquitoes, are only recovered with the greatest difficulty even in areas where other sampling methods such as biting catch, light trapping, etc., indicate the existence of large adult populations.

CHAPTER 4

Mosquito Host Selection and Feeding Patterns

i. Introduction

The blood-feeding contact between mosquito and man is a key factor in determining the role of mosquitoes as vectors of human disease. However, very few species of mosquito feed exclusively on man as the sole host, and the presence of other potential mammalian hosts, such as domestic animals, tends to deviate them from man to a varying degree. This deviation may be accentuated if those mosquitoes which are attracted to man in the first instance are repelled or prevented from feeding by clothing, protective screening or netting or by repellents.

The successful transmission of some mosquito-borne diseases demands a higher and more frequent degree of biting contact with man than others. With malaria, urban filariasis, urban yellow fever and dengue, for example, the mosquito can only acquire its infection in the first instance by feeding on an infective human subject. The pathogens involved in these three diseases all undergo changes, development or multiplication in the mosquito host before that mosquito can become infective in turn — normally after a

51

period of 10-12 days — capable of passing on the infection to a new host in the act of biting or feeding. If the mosquito feeds indiscriminately on other mammalian hosts as well as man, it may fail completely to acquire an infection in its lifetime. If it does have infrequent contact with man, the occasional human host may not be in an infective stage at the time of biting. If the mosquito does manage to acquire an infection by feeding on an infective host, there is the possibility that the mosquito may not live long enough to become infective, or if it does survive it may still feed on other mammalian hosts unaffected by the human malaria parasite or other pathogen in question. In view of these possible dead ends, the successful transmission of such pathogens is absolutely dependent on the mosquito taking a minimum of two human blood feeds on different hosts in the course of its life, these feeds being separated by an interval long enough for infections acquired in the first blood feed to develop to the infective stage in the mosquito.

In the case of those human diseases in which the normal reservoir of infection is an animal (mammal or bird) other than man, such diseases being classed as zoonoses, the minimum two blood feeds on man is not essential to secure transmission. The mosquito vectors of such zoonoses normally feed on mammals or birds and acquire their infections from them. Once the infection has developed to an infective stage, a single feed on a human host — perhaps the only human blood feed in the mosquito's life time — is sufficient to transmit the disease pathogen. In such cases mosquitoes which feed indiscriminately on animals and man are more likely on the one hand to acquire infection from a particular mammal or bird, and also more likely to have sufficient contact with man to enable the pathogen to be transmitted.

The majority of mosquito-borne zoonoses are viral diseases, including Jungle Yellow Fever and various Encephalitides such as Western Equine Encephalitis (WEE), Eastern Equine Encephalitis (EEE), St. Louis Encephalitis (SLE), etc.

In studying the feeding patterns and host selection of mosquitoes, the emphasis and the approach to the problem tends to differ according to which of the two categories above is involved. With malaria, urban filariasis, urban yellow fever and dengue, for example, the main interest is to establish the relative attraction of mosquito to man versus his domestic animals (Cow, horse, sheep, dog, etc.) and the extent to which these animals provide alternative sources of blood meal. In the case of the zoonoses group, the alternative hosts to be considered are not so much domestic animals, but wild or feral animals which are the natural reservoirs of those diseases, and the animals on which the mosquito must feed in the first instance in order to acquire infection.

The pioneer workers on malaria and anopheline mosquitoes very early noted that some species of mosquito feed regularly on man, while others were usually more attracted to cattle or other domestic animals. The extreme examples of these two different categories were classified as anthrophophilic and zoophilic. In general highly zoophilic species, even though they fed occasionally on man, played little part in malaria transmission, and in most regions of the world malaria was found to be naturally transmitted by only one or two of the local species of *Anopheles*. These main malaria vectors were characterized by a high degree of blood feeding on human hosts, and by their associated ability to maintain endemic malaria at comparatively low mosquito densities. However, in many countries malaria is transmitted by anopheline mosquitoes which are more catholic in their host preferences, feeding more indiscriminately on man and on his domestic animals. Such species can nevertheless maintain a high degree of malaria transmission because their particular breeding habits and habitats enable enormous mosquito populations to be produced, thus greatly increasing the likelihood of the human host being bitten. For example, *Anopheles culicifacies* in the plains of India and in Sri Lanka, *Anopheles pharoensis* in the Nile Delta,

Anopheles albimanus in Central America and the Caribbean, and *Anopheles aquasalis* in Trinidad and the north-east coast of South America have in common this liability to occur at high densities. The consequent increased feeding activity on mammalian hosts including man more than compensates for their indiscriminate preferences, and enables those species to be principal vectors of malaria in their respective regions.

According to differences in emphasis discussed above, studies on mosquito host preferences have centred on one hand on man versus domestic animals in the village environment, and on the other hand on man versus wild mammals or birds as alternative hosts in the wild or feral environment. The study methods used can be conveniently grouped under the following headings:

1. Captures of mosquitoes landing/biting on man and on tethered domestic animals in the village environment.

2. The use of the precipitin test for identifying the source of blood in the stomach of engorged mosquitoes.

3. Comparison of mosquito captures in traps baited with different feral mammals and birds, in many cases with the object of studying host preference between such animals rather than as a direct comparison with man.

4. Experimental studies on mosquitoes offered a choice of host in traps or in cages, involving both known and unknown numbers of mosquitoes.

5. Combinations of two or more of the above.

ii.   Nocturnal Captures of Biting Mosquitoes

Captures of mosquitoes on tethered domestic animals has long been used as a basis for determining host preferences of anopheline mosquitoes, and is now widely used in the study of other types of mosquito as well. A particularly instructive example of this study method is the work done in the North Arcot district of Madras State, India, following the original outbreak of Japanese encephalitis in 1955, in which culicine mosquitoes of the *Culex vishnui* group were implicated (Reuben, 1971a-d). Simultaneous all-night mosquito collections were made under trees in a clearing at the village edge on various hosts: a small bullock, a young buffalo of similar size, and two men (Reuben, 1971a). Two teams of four men — one to each animal bait — and two men acting as human bait, worked in 3-4 hour shifts from 1700 to 0800 hours. Collectors worked on a rotation roster, changing places at the end of each hour, so that each man spent an approximately equal amount of time collecting from each kind of bait. Mosquitoes were captured with the aid of an electric torch and an aspirator on the cattle, and directly into test tubes by the human bait collecting from their own and each others' arms and legs. The results of catches taken in 68 all-night collections are shown in Table 4.1 and illustrate well the range of host preference. Only one of all these mosquito species (viz. the domestic *Culex pipiens fatigans*, vector of urban filariasis) could be considered as being attracted almost exclusively to man. All the others, including the vector of Japanese encephalitis, were taken in much greater numbers on the animal hosts. At the same time it should be noted that, with a single exception, all the species did attack man at one time or another, and none of them showed complete deviation to the animals hosts. In general, differences between the attraction of bullock versus buffalo were much less marked than between either of these animals and man, but some species did show a significant preference for bullock over buffalo.

Table 4.1   Mosquitoes caught biting different kinds of bait in the open
(68 night collections), North Arcot, Madras.   (Reuben 1971).

	2 Men	Bait Bullock	Buffalo
Culex p. fatigans	20	1	2
C. bitaeniorhynchus	2279	2496	2303
C. vishnui	2545	66,219	41,353
C. fuscocephalus	35	1566	2259
Anopheles subpictus	21	14,973	6376

Table 4.2   Number of mosquitoes caught biting human bait in three
different situations, and a bullock in the open on 62 nights in a village in
North Arcot district, Madras State.   (Reuben 1971).

	Bait and location			
	2 Men indoors	2 Men in porch	2 Men outdoors	Bullock outdoors
Culex bitaeniorhynchus	6	6	48	232
Culex vishnui	167	239	546	16,777
Anopheles subpictus	12	18	15	12,821

These painstaking investigations in South India, as well as in Pakistan
(Reisen and Aslamkhan, 1978), also provided a mass of information about
biting cycles of the more abundant culicine and anopheline mosquitoes, using
the more generally preferred host, bullock, as the standard bait.   This
aspect of the work, modelled on the classic mosquito biting cycles studies in
East Africa, will be discussed in a different context later.

Using the relative attraction of two men versus the preferred animal
host (bullock), studies were also carried out in another village on the
effect of location on biting preference.   Collections on two men indoors were
compared with two men in the porch, two men outdoors and the bullock
tethered outdoors.   Table 4.2 shows that most mosquitoes were taken in larger
numbers on the bullock in the open.   Next in order of preference were the
two men in the open, followed by two men in the porch.   The two men inside
the hut attracted least of all — apart from a small collection of *Anopheles
subpictus*.   This study revealed that the virus vector, *Culex vishnui*, despite
its general preference for animal hosts, will still feed indoors on man to a
considerable extent when alternative hosts, men and animals, are available
outdoors.   At the time of this study the villagers were normally sleeping
indoors, and consequently the mosquito vectors house-entering habits would
favour the chance of a successful feed on the sleeping human host.

iii.   The Precipitin Test for Mosquito Blood Meals

The essential requirement of this technique is to prepare anti-sera
specific for man and the main groups of animals by immunising rabbits with
the appropriate sera.   Of the two groups of antigen produced in the rabbit's
blood, the heterologous portion is removed by adsorption, and it is the
homologous sera which are used for the precipitin test.

The stomach contents of freshly or recently-engorged mosquitoes are expressed onto filter paper, dried carefully, and sent for analysis to a central laboratory. In the laboratory, the blood spots or smears cut out of the filter paper are eluted in saline, and then — by means of a small capillary tube — brought into contact with the specific antisera. If the insect has fed on the host for which the test is being made, a white precipitate develops at the interface. The accuracy of the test depends on the use of antisera of high titre, specific for the animal or group of animals. The anti-sera commonly used in the laboratory react with dilutions of the homologous serum of 1:128,000 or greater.

The precipitin test is commonly used to distinguish the blood of man from that of associated domestic animals in the mosquito blood meal, these being commonly cow/ox, horse/donkey, pig, sheep/goat, and dog. The test cannot differentiate between closely related animals, such as between man and primates, sheep and goat, horse and donkey, wild and domestic pig, etc. Where hosts other than domestic animals are involved, as occurs commonly with tsetse flies, sandflies, triatomine bugs, etc., a more specific but at the same time more elaborate agglutination test has been developed. These and other refinements of the basic precipitin test will be dealt with in more detail later in the section.

In one of the earliest applications of the precipitin test for identifying the blood meals of wild-caught anopheline mosquitoes in malaria surveys, the field worker was provided with samples of anti-sera and a simple do-it-yourself kit suitable for distinguishing human blood from that of domestic animals. In many cases the essential information required was the proportion of engorged mosquitoes which had fed on man versus cattle, but the test could also provide facilities for identifying the blood of horse, sheep and pig. Simple field kits of this kind played an important part in some of the classical studies on malaria vectors, such as those of the 'biological races' of *Anopheles maculipennis* in Europe in the 'thirties.

However, despite the great advantages to the field worker of being able to identify his blood smears quickly on the spot and obtain immediate answers, there were disadvantages in the way of accuracy and uniformity. In all blood meal series, a variable proportion were found to be negative to all five main groups of animal tested, and there was doubt as to whether these blood samples came from other sources, or whether the blood sera had failed to react because of its age, insufficient blood proteins or some other reason.

A great step in progress occurred in 1955 when the World Health Organization in agreement with the Lister Institute of Preventive Medicine, England, set up a precipitin testing service related to entomological surveys in malaria eradication programmes, the service being available to national research workers as well as to WHO field staff. The main object was to intensify the study of blood-feeding patterns of anopheline mosquitoes, and to eliminate experimental errors by using standardized techniques and highly sensitive antisera. By the end of the first 5 years of this operation, over 56,000 tests on 51 species of anopheline had been carried out, giving 94% positive results. By the end of 10 years, the number of tests had reached 124,000, and extended to 92 different species or species complexes (Bruce-Chwatt and co-workers, 1966).

Since 1967 the Imperial College Field Station in England has provided the test centre for identification of arthropod blood meals, not only for mosquitoes but for a wide range of blood-sucking arthropods from all over the world.

The use of the precipitin test as described so far is adequate for most studies on anopheline vectors of malaria where man and domestic animals provide the main choice of hosts in and around human settlements. A rather different range of problems is postulated by arboviral diseases transmitted by culicine mosquitoes where man and domestic animals are only incidental hosts.

It became important in such cases to identify blood meals from a much wider range of natural or potential hosts, including those which provide the main reservoir of infection in different zoonoses.

Among these alternative hosts, birds urgently required investigation in view of the fact that they are the natural hosts of many species of *Culex*. No information about this was available from the extensive precipitin testing programme described above and coordinated by the WHO, as prior to 1960 the inability to produce antisera with sufficient specificity within the avian groups resulted in a dearth of data about the possible role of birds as alternative or preferred hosts of mosquitoes.

In order to improve specificity within the avian group, the chicken was used as the antibody producer in preference to the rabbit. The resultant antiserum enabled blood meals obtained from birds to be classified according to orders. The use of chicken-produced antibodies also improved the sensitivity with regard to mammalian blood (Tempelis, 1975).

The availability of such a valuable research tool as the precipitin test would appear to offer unique opportunities for establishing once and for all the relative degree of feeding contact between different species of mosquito and man. This would undoubtedly be true if one could be sure that the samples of mosquitoes collected for blood smears were completely representative of the whole feeding population. But in practice this ideal is very difficult to attain. Earlier workers were fully aware that the samples of mosquitoes collected for blood smears were heavily biased according to the collecting site. Samples of engorged mosquitoes collected in human habitations showed an obvious bias for human blood, while the predominantly stable or cowshed samples showed mainly domestic animal blood.

Quite apart from the main objective of establishing the feeding patterns of proved or suspect malaria vectors, the precipitin tests provided the most convincing proof of mosquito movements after blood feeding. Not only were mosquitoes containing human blood frequently taken in nearby animal shelters, but mosquitoes with ox blood were also found among the predominantly human blood to be expected in exclusively human dwellings and sleeping quarters.

The precipitin tests have also provided convincing proof that mosquitoes can feed on more than one type of host in the course of engorgement, and there are several records of single blood smears containing the blood of man as well as that of some other domestic animal such as ox or goat (Boreham and Lenahan, 1976).

To return to the main objective of precipitin testing, there are still many difficulties in interpreting the results of the massive series of tests collated by the WHO. From the epidemiological point of view, these figures — taken in conjunction with the record of mosquito infection rates with malaria parasites (oocysts in the stomach and sporozoites in the salivary glands) — have confirmed the role of certain species of *Anopheles* as important vectors of human malaria. On the basis of precipitin test results, anopheline mosquitoes have been arbitrarily divided into three categories according to the human blood ratio, i.e. the proportion of positive blood samples reacting to human serum, namely, those with the ratio less than 10%, those between 10% and 50%, and those with over 50% (Bruce-Chwatt and co-workers, 1966). Despite one or two species whose human blood ratio would appear to be too low to be in accord with their established role as malaria vectors, the major vectors of malaria are, in general, also those with a consistently high human blood ratio.

Some species of anopheline have been repeatedly found with sporozoite infections under conditions in which they do not appear to be involved in human malaria transmission, and consequently some doubt has arisen about the exact identity of the sporozoites. If in such cases the precipitin test record shows no or minimal blood feeding on man, then the chances are that the malarial infections in the mosquito are very likely of non-human origin.

It will be seen, therefore, that one of the major difficulties in the way of correctly interpreting precipitin test records of anopheline blood meals in particular and mosquito blood meals in general is the obvious bias caused by collecting the mosquito sample from a population under conditions in which a particular host predomonates. In order to counteract this bias many investigators have carried out tests on engorged mosquitoes collected in a variety of outdoor resting places, both natural and artificial. With some species this has only been possible to a limited extent because of inadequate knowledge of the outdoor day-time resting sites, and consequent difficulty in obtaining samples comparable to those taken in human habitations or animal shelters.

In recent years the interpretation of mosquito blood meal records in terms of defining normal feeding patterns has been further complicated by two factors. Firstly, the development and intensification of the WHO-coordinated precipitin-testing programme coincided with the period of greatest intensification in the global malaria eradication programme based essentially on treatment of all indoor resting sites of mosquitoes with DDT and other residual insecticides. Since then, many of the mosquito populations sampled have been under direct insecticide pressure or have been affected indirectly by country-wide programmes in such a way as to produce drastic changes in the proportion of engorged mosquitoes resting indoors versus outdoors, or even more significant changes in their natural feeding patterns.

The second development which has greatly affected the interpretation of precipitin test data is the comparatively recent clarification of species complexes such as that of *Anopheles gambiae*, the main malaria vector of tropical Africa. It is sufficient for the moment to say that *An. gambiae* is now known to be a complex of at least four different forms which were indistinguishable at the time of maximum precipitin testing activity. As at least two — sometimes three — of these forms occur together in most parts of Africa, and as these types are now known to differ in feeding and resting habits, it is difficult — if not impossible — to say what the composition of the original samples of '*Anopheles gambiae*' were with regard to species A, B and C. This problem will be discussed in greater detail later in this section.

Feeding patterns of culicine mosquitoes. The application of precipitin testing to culicine mosquitoes — particularly those species established as vectors or arbovirus disease — was slow to develop, and for many years received scant attention compared with the dominant interest in anophelines and malaria.

However, since about 1960 the question of blood feeding patterns of culicine mosquitoes has been intensively studied by research workers at the School of Public Health, Berkeley, California (Tempelis, 1975; Tempelis and Reeves, 1964; Tempelis and Washino, 1967; Tempelis and co-workers, 1970), and later by other US workers in Florida (Edman and co-workers, 1972; Edman, 1974, 1979a,b) and elsewhere (Hayes and co-workers, 1973; Washino and Tempelis, 1967). Although the main object of this work was to provide much-needed information about culicines, the investigations have also included, or been extended to, US species of anophelines. The advances in that field too have provided new insight into many long-standing problems and puzzles regarding the main malaria vector species in the world.

The whole approach to this blood-feeding problem by the US workers has been on such a sound and scientific basis that it now provides a simulating model and guidance for future work. The outstanding features of this work are as follows. Firstly, more sensitive serological techniques were developed for the identification of many blood meal sources, especially in the case of the long-neglected birds. By using chicken to produce the antisera it was now possible to identify blood samples from different orders of avian host, and to identify different species in the case of gallinaceous birds. For detecting the blood of the domestic chicken itself, an antiserum was produced in the pheasant.   These improved the more sensitive techniques have also permitted more accurate identification of mammalian blood meals.

Secondly, from the start of these investigastions the aim has been to obtain the necessary blood smears or stomach contents from mosquito samples collected from a wide range of resting sites.   Engorged mosquitoes have been collected from such places as farmyard sheds, abandoned animal shelters, house porches, artificial shelters, culverts, etc.   In some investigations additional samples were obtained from vegetation by means of powered aspirators, and also from light trap collections.

The third point which has repeatedly been emphasized in these feeding pattern studies is the fact that choice of host by mosquitoes is undoubtedly affected by the availability of different animals (mainly mammals and birds). Workers in the malaria field have long been aware of the importance of the man:domestic animal ratio, and there are several records showing a direct relationship between increasing number of cattle relative to man, and decreasing proportion of blood means positive for man.   The US studies have been concerned not simply with the man:cattle ratio, but with a much wider range of mammmalian and avian hosts.   These ideas have been crystallised in the form of the 'Forage Ratio', which is the percentage of engorged mosquitoes which have fed upon a given vertebrate host divided by the percentage which that host comprises of the total population of hosts available (Hess and co-workers, 1968).   A forage raio of 1 or near 1 suggests indiscriminate choice of hosts;   forage ratios significantly greater than 1 indicate selective preference, while ratios significantly less and 1 indicate avoidance in favour of other hosts.

The value of the forage ratio approach is well illustrated by studies on two common culicine mosquitoes on the Island of Oahu, Hawaii, viz. *Culex quinquefasciatus* and *Aedes albopictus* (Tempelis and co-workers, 1970).   If one assesses the feeding preference of these two mosquitoes by results of precipitin tests alone, they both appear to feed mainly on birds.   However, a census of availability of mammals and birds in the different study areas showed that while the feeding of *Culex quinquefasciatus* was in line with the relative availability of birds, that of *Aedes albopictus* showed a forage ratio for mammals over ten times that for birds, revealing a clear preference for mammalian blood when mammals — even in low numbers — were available.

In one of the study areas, a dairy farm with over a hundred head of cattle and an abundance of other domestic animals including horses, dogs, chickens, pigeons, ducks and geese, *C. quinquefasciatus* showed a 3:1 preference for birds over mammals due primarily to its high forage ratio for chicken (3.5) and low forage ratio for cows.   This difference is even more impressive when we consider the much greater size of the cattle, and the fact that they comprised nearly half of the total available domestic animal hosts.

This study with the extended range of precipitin sera also revealed a marked selection among the avian hosts themselves on the part of *C. quinquefasciatus*.   The forage ratio for chicken was 35 times that of ducks, even though these two species were comparable in size and would appear from casual observation to be equally suitable sources of blood meal.

It is clear from this work that one cannot make valid interpretations of the precipitin test record in terms of host preference unless data are available on the relative numbers of different potential hosts in the study area.   In view of the fact that in some areas there are real difficulties in obtaining accurate counts or censuses of available vertebrates, rough population estimates have to suffice.   Nevertheless, in many cases crude estimates may be sufficient to indicate trends and provide some information about host availability without which the interpretation of the precipitin test record becomes extremely difficult.

With the accumulation of accurate information from different parts of the

United States about the identity of mosquito blood meals, confirmation was obtained that while most *Anopheles* and *Aedes* feed almost entirely on mammals, birds form the most important source of blood meal for a wide range of culicine mosquitoes other than *Aedes*. Of particular interest from the virus-transmission point of view were those species which make use of both avian and mammalian hosts. The common and widespread *Culex pipiens quinquefasciatus* feeds readily on both birds and mammals; *Culex erythrothorax* feeds almost exclusively on mammals in one geographical area and on birds in another.

One of the most significant revelations was that two of the most important arbovirus vectors, *Culex tarsalis* — vector of Western Equine Encephalitis (WEE) in California (Tempelis and Washino, 1967), and *Culex nigripalpus* — vector of St. Louis Encephalitis (SLE) in Florida (Edman and co-workers, 1972) — fed preferentially on birds in the spring and early summer but showed a shift to mammalian hosts in the middle and late summer season. This change seemed to be influenced by different factors in each case. In California, the great increase in *Culex tarsalis* density in summer may be more than the available bird population can tolerate, hence the overspill to the more easily accessible mammalian hosts. By following the changes in mosquito population by means of light traps, it was shown that the height of mammalian feeding occurred in the month of peak *C. tarsalis* production, in California and in Texas.

In *Culex nigripalpus* in Florida, on the other hand, the shift from complete avian hosts in winter and spring to equal or near equal avian and mammalian hosts in summer is not entirely related to high summer mosquito densities. The climatic changes at that season, particularly the increased humidity, stimulate many mosquitoes to leave their woodland daytime resting places and fly out into open fields where mammalian hosts are more likely to be encountered.

These findings with regard to the seasonal shift in feeding pattern mark an important advance towards a better understanding of the complex epidemiology of arbovirus transmission. The seasonal change in host preference provides the mechanism whereby the viruses of WEE and SLE in their avian reservoirs can be transmitted to the susceptible host — man or susceptible domestic animals — and thus produce sudden outbreaks or epidemics of those diseases.

Of equal importance, particularly in the context of the present theme, is the way in which these findings based on greatly improved precipitin testing and its application have stimulated a more critical approach to experimental studies into why and under what conditions some hosts are preferred to others. This can best be presented in the following section alongside a description of other experimental methods used in mosquito studies in general (see (v), p. 65).

The increasing range and sensitivity of precipitin testing have stimulated further critical studies in other areas outside the United States. In the continuing work in the Gambia, for example, host preference studies have occupied as important a part of the programme as the flight patterns already described (Snow and Boreham, 1973, 1978). Twenty-three species of *Aedes* and *Culex* were represented in extensive precipitin testing of blood meals of mosquitoes taken from four sources: (i) resting on the outside of screens and walls of houses; (ii) unbaited suction traps in woodland, regenerated bush and fallow fields; (iii) unbaited suction traps in mangrove and adjacent tidal flats; and (iv) in the course of house catches.

Three main types of feeding pattern were disclosed: those in which large mammals were the predominant hosts (e.g. *Culex tritaeniorhynchus*), those in which mammals, birds and reptiles were represented (e.g. *Culex thalassius*), and those in which avian feeds predominated (e.g. several species of *Culex*). The Gambian precipitin test surveys are very illuminating, not so much from the results actually reported, but in that they illustrate how difficult it is to eliminate all possible sources of error in applying this technique and in interpreting the results. For example, samples of mosquitoes from unbaited suction traps would appear to be un-biased from the point of view of hosts available, but in practice the presence of people attending the suction traps made it difficult to say if records of primate blood were those of man or of local monkeys. Although the overall picture is one of varied and representative sampling, each species tended to be abundant in one collecting site and scarce in others. In many cases the dominant collecting sites were associated with human and animal settlements, as for example with over 1000 smears of *Culex thalassius*, 92% of which came from females resting on the wall of a screened bungalow. Particularly interesting too is the very small number of engorged *Aedes* taken in these extensive surveys (a total of 19 representing 7 species), the five *Aedes aegypti* all being taken in or on the outside of human habitations.

The Gambian investigations also reveal the great difficulty in applying any sort of forage ratio in such areas where discrete small areas of irrig-ated rice are visited by birds from distant roosting sites. Many of the birds present by day in the study area, and thus available for census, depart for distant roosts at dusk and thus become unavailable as hosts for local night-biting mosquitoes. However, the tests did reveal feeding on a wide range of avian groups, and that while some species of *Culex* fed predominantly on passerines other species fed on a wide range of avian families.

Fortunately, in the Gambian studies, precipitin testing was not the sole method used to establish feeding patterns and host preferences. The use of traps baited with different potential hosts was extensively used, as will be described in section vii.

The increasing sensitivity and host range of precipitin testing has also been fully utilized in East Africa (Chandler and co-workers, 1975, 1976, 1977). Following the decision of the Kenya Government to irrigate some tens of thousands of hectares of the Kano Plain in Nyanza Province, a pilot irrigation scheme of 810 ha was completed in 1968. In the following year intensive studies were initiated as part of the Medical Research Council's overseas viral research programme. The precipitin test survey was carried out in an area of irrigated rice fields with adjacent heronries and a large resident bird population. Mosquitoes were collected from outdoor sites using battery-operated hand aspirators, plus collections made in long grass and vegetation beneath the trees in which the birds were nesting. Bird blood meals could be identified to Galliformes (chicken), Anseriformes (ducks), Ciconiiformes (herons, egrets), Columbiformes (doves, pigeons), Falciformes (eagles, kites) and Passeriformes (weavers). Tests showed that *Culex univittatus*, known to be an important vector of Sindbis and West Nile virus in Africa, and *C. poecilipes* had high preferences for avian blood, mainly of herons and egrets. Another species of *Culex*, (*C. antennatus*) and two species of *Mansonia* only occasionally fed on birds.

Further studies in the Kisumu heronry in Kenya provided a fresh insight into seasonal changes in blood feeding patterns. Mosquito collections were made throughout a complete year in the heronry, and in addition regular notes were made of nesting bird populations. Although it was originally concluded that *Culex univittatus* was mainly ornithophilic, the recovery of considerable numbers of females containing rodent blood early in this special investigation prompted regular recording of rodent populations as well, in particular the grass mouse (*Arvicanthis abyssinicus*), by means of live-trapping techniques. Outdoor collections of mosquitoes were made twice weekly in long grass and bushy vegetation beneath trees in which birds were nesting. The blood meal from gorged females was expressed onto filter paper and sent to the UK for testing against order-specific antisera, enabling six groups of birds to be distinguished, viz. chickens, herons and egrets (Ciconiiformes), ducks, doves and pigeons, eagles and kites, and weavers.

Of the large number of different mosquito species captured, only the dominant *Culex univittatus*, *C. antennatus* and *C. circumluteolus* were caught in sufficient numbers to allow an analysis of seasonal variation in feeding. The results of an impressive series of precipitin tests (2051 in the case of *C. univittatus*) revealed not so much any clearly defined change in host preference but rather a flexibility in feeding habits according to the available hosts and also according to the mosquito density at the time. The number of *C. univittatus* increased during the 'long rains' of March to May, the high peaks being accompanied by the highest proportion engorged. During these months of high mosquito density, Ciconiiform birds (herons, egrets and others) nested in the heronry, and their high availability to mosquitoes is reflected in an increase in avian feeds at that period.

In the months when these birds were absent from the heronry, numbers of avian-fed *C. univittatus* were relatively low and there was evidence of a shift to other groups of birds such as ducks and weavers. Although there was a positive correlation between the number of ciconiiform birds at the heronry and the percentage of *C. univittatus* which had fed on these birds, beyond a certain number of birds available — in this case about 500 — the percentage of ciconiiform blood did not increase further but levelled off at approximately 75%.

A similar opportunistic feeding pattern is suggested by the high proportion (23%) of rodent blood samples recorded in the dry months at the beginning of the investigation, which could be attributed to scarcity of alternative hosts such as grazing cattle and nesting birds. When rodents were offered in baited traps during the breeding season of the herons they were fed upon to a very limited extent, a further indication that when nesting birds are present they are the principal hosts for hungry *Culex univittatus*.

While the precipitin test remains the main bulwark for the identification of blood meals of mosquitoes and other blood-sucking flies, it has long been recognized that the test is unable to distinguish between the blood of closely related animals such as sheep and goats, and man from other primates. It now appears that some of the deficiencies may be dealt with by using an entirely different method of blood-meal identification. This is the haemoglobin crystallization which utilizes the fact that the crystal structure may show distinctive features according to the animal source (Washino and Else, 1972). Some closely related species can be differentiated in this way. In the Bovidae, for example, the crystal patterns for cow, sheep and goat are distinctly different. In addition the crystal structure is still identifiable for a much longer period (up to 36 hours) after the mosquito's blood meal than is normally possible with the precipitin test.

The possibilities and the practical value of this new technique are well illustrated by the latest studies carried out in Florida on the blood feeding habits of culicine mosquitoes (Edman, 1979a,b). There, the haemoglobin crystallization test was used to supplement the routine precipitin tests specifically for distinguishing different rodent blood meals. Preliminary tests with whole blood samples and with known mosquito blood meals showed that raccoon, opposum, woodrat and cotton rat always produced numerous and easily recognizable specific forms of haemoglobin crystals. Rice rat and cotton mouse, however, proved to be less satisfactorily distinguishable.

Tests carried out on 138 blood meals of *Culex opisthopus* captured in nature showed that 110 (80%) produced specific crystal forms. 54% of these positive blood types belonged to rodents of which it could be demonstrated that cotton mouse and cotton rat were the major blood types found. *Culex opisthopus* is a member of a subgenus, *Melanoconion*, among which it is apparently unique in being principally a mammal feeder. It is also unique among mammal-feeding mosquitoes in that area of Florida in the high proportion feeding on rodents. In this case the specific rodent hosts of *Culex opisthopus* were those which were most common, and most frequently incriminated, in the cycle of Venezuelan Equine Encephalitis (VEE) in southern Florida.

iv.   Mosquitoes Attracted to Animal and Bird-Baited Traps

The blood-feeding habits of mosquitoes involved in the transmission of virus between different animal hosts and between those hosts and man is a major problem for study in many parts of the world where such infections are liable to be of public health importance.   Rather than attempt to do justice to all the work that has been done, or to attempt to summarize the great mass of information about so many different species of mosquito concerned, it is felt more instructive to select a few more recent illuminating examples which illustrate the different ways in which fundamentally similar problems have been tackled.   All of these recent studies have been influenced, and to some extent guided, by the earlier investigations in Central Africa and South America in the 'forties and 'fifties.   Those pioneer research projects laid a solid foundation for future work, but they will be omitted from this present discussion as they have already been well documented, summarized and reviewed elsewhere.

One of the most active research centres in recent years has been the Arbovirus Research Unit associated with the South African Institute of Medical Research (Jupp, 1978;   Jupp and McIntosh, 1967, 1970;   McIntosh and co-workers, 1972, 1977).   Entomological studies have been carried out in two main field bases: firstly in the Ndumu Game Park on the coastal region of North Natal along the Mozambique border, where the main subject has been the study of mosquitoes transmitting Chikungunya (CHIK) virus from the natural host, the Vervet Monkey (*Cercopithecus aethiops*), to man.   The second base is on the inland plateau, or highveld, on the outskirts of Johannesburg, where studies have been concentrated on transmission of Sindbis and West Nile Virus infections of man.

In the Game Park, man, monkey and fowl were used as baits for mosquito collections, all of which were made simultaneously at ground level and in the canopy (12 m) in gallery forest where monkeys had previously been found infected with Chikungunya virus.   The man bait collections at ground level and on the canopy platform were made by two people collecting mosquitoes as they landed on exposed legs.   Catches started half an hour after sunset and ended $1\frac{1}{2}$ to $2\frac{1}{2}$ hours later.   In each fowl-baited trap one fowl, restrained within a nylon stocking, was used in a black circular trap of the 'lard can' type previously used in American work, and consisted of a tube 66 cm long and 33 cm diameter with an inverted mosquito gauze cone at each end, provided with a small central hole.   Monkey baited traps were of similar design, but larger, and the monkey was restrained within a wire cage.   Observations were carried out during the six wet months of the year, and involved over 30 species of mosquito.   In addition to the capture data normally recorded in such collections, a 'persistence index' was calculated for each mosquito with each class of bait.   This is the proportion, expressed as a percentage, of the days on which the species was collected to the total number of collecting days.   It was found to be a useful measure of the relative time each species was encountered as a blood-seeking female.

In interpreting the results of these catches, the South African workers were acutely aware of possible limitations and snags in their techniques. For example, several species of mosquito known to feed readily on monkey in the laboratory failed to be taken — or only in small numbers — in the monkey traps.   It therefore seemed likely that for some species the nature of the trap, or the obstruction of the wire cage or mesh, acted as a deterrent.   It will be recalled that difficulties of this kind with monkey-baited traps had been encountered several years previously in viral studies based on Entebbe in Uganda. This finding also introduced some uncertainty into the fowl trap collections, where low entry by some species could be due to low attraction to the bait or to some deterrent effect of the trap. The main suspect vector species, mosquitoes of the *Aedes furcifer/taylori* group, were readily taken on man on the platform, but were among those mosquitoes reluctant to enter monkey-baited traps.

Despite the range of bait provided, only about one-third of the known mosquito fauna of that area were recorded in this survey, no information becoming available about the feeding habits of the bulk of the mosquito species. Among the most noteworthy absentees was *Aedes aegypti*, whose immature forms were extremely common in tree holes in the forest. Its non-appearance could either be attributed to feeding preferences other than those provided, or its peak of biting activity may have occurred at a time of the day or night when routine collections were not made.

Following a fresh outbreak of Chikungunya fever, a new monkey-baited trap was designed without the previous restrictions (Jupp, 1978). A young baboon (*Papio ursinus*) or Vervet monkey (*Cercopithecus aethiops*) was anaesthetized and roped to a bait platform, either at 1 m or at 10 m above the ground. Under the platform two rubber-bladed suction fans in a cylinder sucked into collecting cages any mosquitoes approaching the bait (Fig. 4.1). In this case large numbers of the *Aedes furcifer/taylori* group were collected at both levels, unbaited control traps giving zero catches.

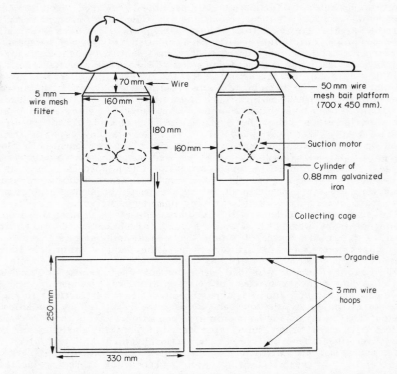

Fig. 4.1 Diagram of trap for collecting mosquitoes attracted to monkeys and baboons. (After Jupp, 1978).

In the ecological studies in the highveld, the main mosquito vector of Sindbis and West Nile virus is *Culex univittatus* (Jupp and McIntosh, 1967, 1970). A variety of mammal and bird-baited traps were again used, including the lard-can trap which was known to be an adequate baited trap for three species of culicine, namely *C. pipiens*, *C. theileri* and *C. univittatus*, but not for others. Particular emphasis was given to the baited net trap, a technique used successfully in other parts of Africa as well as other countries. This consists simply of a large mosquito net, 7 x 5 x 6 ft high set on poles with the bottom edge raised 8 inches from the ground all round so as to allow free access to hungry mosquitoes attracted to bait inside the net. The net is let down at intervals to allow the mosquitoes inside to be captured. This trap has the advantage of being operable all night.

This series of tests, which included several wild bird species as bait, enabled a general blood preference picture to emerge. *Culex univittatus*, apparently the primary wild cycle vector of both viruses, feeds mainly on wild birds and on domestic fowls. However, it has a low feeding preference for man and seems unlikely to be the main vector causing human infection. Some, if not most, of the transmission to humans is probably caused by *Culex theileri* and/or *Aedes lineatopennis*, both of which feed significantly on birds as well as on man. In addition they will also feed indoors, making successful feeding on man more likely.

For many years Trinidad has been the centre of studies on the ecology of two important New World viruses, namely Venezuelan Equine Encephalitis (VEE) and Eastern Equine Encephalitis (EEE). A ten-year study from 1953 to 1963 by the Rockefeller Foundation on possible arthropod vectors of these diseases had recorded frequent isolations of VEE from *Culex portesi*, while EEE was isolated four times from *Culex taeniopus*. The association of each of these mosquitoes with the transmission of different viruses suggests that the two species might have different host preferences, which might in turn reveal the most likely animal reservoirs for VEE and EEE respectively.

In the Trinidad swamp forest where *C. portesi* and *C. taeniopus* occur, the greater part of the vertebrate fauna is limited to lizards, bats, rodents and marsupials (Davies, 1978). The fact that most of these animals are smaller than a rabbit provided a unique opportunity for studying the attraction of animals which could be easily handled, and could be contained in one size of trap. For this purpose, a very satisfactory type of trap was designed on the suction principle. The propellor component of this was activated by a time switch so that it operated the fans for 45 seconds every $7\frac{1}{2}$ minutes. This interval allowed mosquitoes to approach the bait cage suspended below the trap and for their numbers to build up for $7\frac{1}{2}$ minutes at which time the air current produced by the propellor sucked the mosquitoes into the trap, egress from which was prevented by a valve.

The trap was suspended with the bait cage 90 cm above the ground so that adjacent traps were 13.5 m apart around the circumference of a circle of radius 13.5 m, at the centre of which the switch and batteries were located. All bait cages were made as similar as possible, and were bleach sterilized after every night to eliminate residual odours.

The range of hosts provided consisted of one species of bat, eleven species of rodent, three species of marsupial, two species of bird, one reptile, one amphibian and one crustacean. The 'attraction' of a bait animal was estimated by the number of mosquitoes collected by a suction trap containing that bait during one night's operation. This implies that the mosquitoes 'attracted' must have approached to within about 15 cm of the bait, and remained there until the next suction cycle.

Not every mosquito attracted to the bait managed to take a blood meal. The chances of successful feeding are governed by three main factors: firstly, the availability of the animal in the mosquito's habitat; secondly, the attraction exerted by the animal; and thirdly, the availability of blood to the mosquito including various physical barriers. By combining figures

Table 4.3   Eighteen bait animals ranked according to engorgement index
for both species of mosquito.   (After Davies, 1978).

Bait	*Culex portesi* index	Bait	*Culex taeniopus* index
Didelphys	10.67	Didelphys	10.99
Nectomys	8.45	Caluromys	7.22
2 white mice	7.87	2 white mice	4.56
Agouti	6.84	Squirrel	3.55
Guineapig	6.61	Guineapig	2.63
Caluromys	6.34	Marmosa	2.54
1 white mouse	5.88	Hamster	2.22
Hamster	4.68	Agouti	1.61
2 screened white mice	4.33	Nectomys	1.42
Proechimys	4.17	1 white mouse	1.09
Heteromys	3.88	Proechimys	0.82
Oryzomys	3.73	Heteromys	0.81
Squirrel	3.71	Oryzomys	0.71
Akodon	3.51	Dove	0.45
Marmosa	2.92	2 screened white mice	0.32
Rhipidomys	1.31	Rhipidomys, Bat and	
Bat	0.67	Akodon	0.00

for the relative attractiveness of the baits, and the proportion of mosquitoes that could obtain blood from the bait animals, an estimate was arrived at representing the relative likelihood of a mosquito obtaining blood from any of the baits under study.   This was called the 'engorgement index' and was a product of the adjusted mean catch and the percentage of either species of mosquito that engorged at each of the baits.   The results shown in Table 4.3 show that for both species of mosquito the opposum, *Didelphys*, was the animal on which they were most likely to feed.   *Culex portesi* is more likely to engorge on a wider range of baits than *C. taeniopus*.   At the other end of the scale the grass mouse, *Akodon urichii*, ranked high as an attractant for *C. portesi*, but very few of those mosquitoes actually fed on the host.

v.   Cage and Trap Experiments on Host Selection

The extensive precipitin test survey carried out in California on *Culex tarsalis*, vector of Western Equine Encephalitis, and on *C. nigripalpus*, vector of St. Louis Encephalitis in Florida, clearly demonstrated that both species feed mainly on birds in winter and spring, but show a considerable increase in mammal feeding during the summer (page 59).   In California this shift in host preference appeared to be related to increased mosquito population during the summer, while in Florida other factors such as mosquito behaviour changes, caused by climatic conditions in the hot rainy season, appeared a more likely explanation.

These possibilities have been further critically analysed by the use of experimental methods of considerable interest in other similar situations elsewhere.   In Florida the technique was to expose different bait species in large standard mosquito cages, 8 x 8 x 8 ft, into which controlled numbers of unfed *Culex nigripalpus* females could be introduced, and allowed to feed overnight (Edman and co-workers, 1972).   In the first series of experiments using nine different species of ciconiiform (wader) birds, each host was exposed to 300 mosquitoes per cage, this number being chosen after prelim-

inary trials had shown that it was a critical density level. These tests
confirmed that birds may react defensively in various ways to mosquito
attack, and that those species which exhibit intense foot stamping, foot
slapping and foor pecking behaviour were able to prevent the great majority
of mosquitoes from feeding.

In more extended tests, the range of experimental hosts included two
species of bird which had previously been shown to be relatively passive to
mosquito attack (i.e. at 300 per cage), viz. the Black Crown Night Heron
Nycticorax nycticorax, and the Green Heron *Butorides virescens*, and two
which are extremely active, namely the White Ibis *Eudocimus albus* and the
Cattle Egret *Bubulcus ibis*. In the first experiments conducted in the autumn,
three mosquito densities were tested (100, 400 and 1200 females per cage),
each of the four birds being tested at each density for four nights. In the
second experiment carried out during the summer, six more densities were
tested (25, 50, 100, 200, 400 and 800 per cage). The success or otherwise of
feeding was judged on the subsequent recovery rate of caged mosquitoes and
the proportion of these which had blood fed. The results, charted in Figure
4.2, showed that the relatively inactive Night Heron and Green Heron, which
normally allow all the test mosquitoes to feed, became sufficiently active to
prevent a considerable proportion from feeding when the density was greatly
increased. The normally active species, White Ibis and Cattle Egret, dis-
played even more intense antimosquito behaviour at high densities, to such an
extent that at densities of 200-800 only a few more mosquitoes managed to
feed successfully than at the much lower densities of 25-100.

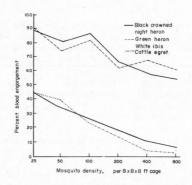

Fig. 4.2   Relationship of *Culex nigripalpus* density to feeding success
(engorgement) on four species of ciconiiform birds.
(After Edman and co-workers, 1972).

Observations had shown that the activity level of the White Ibis was
even higher than that of the Cattle Egret, and its continuous scissoring foot
pecks caused a higher mosquito mortality — over 50% — in all cases.

The implication of all these findings was that with increasing mosquito
density, whether due to seasonal effects or otherwise, a critical level of
tolerance to attack by the normal avian hosts could compel hungry mosquitoes
to seek alternative and more amenable sources of blood, possibly mammalian.

The established seasonal shift in bird:mammal feeding preference has
also been approached experimentally in the case of *Culex tarsalis*, vector of
both WEE and SLE in California (Nelson and co-workers, 1976). In order to
investigate relationships between numbers, feeding success and blood meal
source of *C. tarsalis* baited stable traps were set up with bird or mammal as
as bait. Unlike the Florida experiments, the numbers, age and species of

mosquito in these tests could not be controlled as the mosquitoes entered the traps from field sources. On most occasions, however, *C. tarsalis* comprised about 90% of the total entry.

Bait animals were exposed in pairs, one bird and one mammal per trap, and each two animals were repeatedly exposed together. The mammalian component was the Black Tailed Jack Rabbit, and the avian component either domestic chicken or Ring Necked Pheasant. Animals in cages were exposed overnight, and all mosquitoes collected from inside the trap in the morning. Samples of up to 100 fully engorged females per species were selected for blood meal identification.

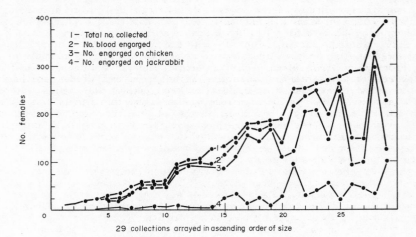

Fig. 4.3 *C. tarsalis* collected in stable traps baited with a jackrabbit and a chicken, and numbers that were blood engorged, fed on the chicken, and fed on the jackrabbit, Llano Seco and Gray Lodge, 1972. (After Nelson and co-workers, 1976).

The results, charted in Figure 4.3, show that there was evidently diversion of mosquitoes from chicken to jackrabbit beyond a critical mosquito density level. Up to collection 16 there was a steady increase in the total number fed, the bulk of this comprising chicken feeds. From that stage onwards the continued increase in blood-feeding was interrupted from time to time by sharp falls in chicken feeding, which were matched by sharp increases in mammalian feeding. These peaks and troughs were particularly marked at the highest mosquito density recorded and showed a trend towards high rates of feeding on the jackrabbit in the larger collections. At this point there was also a sharp overall drop in the proportion of entering mosquitoes which managed to feed successfully.

When pheasant was used as the avian bait, lower engorgement rates and a greater incidence of partly-gorged specimens, combined with high feeding rates on the jackrabbit, indicated that this bird is relatively intolerant or unattractive to *Culex tarsalis*.

The results of these two quite independent studies on factors influencing choice of host, or change or hosts, certainly support each other in under-lining the importance of two main factors, namely density of attacking or biting mosquitoes, and intensity of defensive reaction of the host. Much earlier observations had pointed out that adult pigeons, which are much more active than nestlings, were highly resistant to mosquito attack. The wide-spread practice of restraining avian bait in nylon stockings also recognizes that the restless unrestrained host is less likely to be successfully fed on than an inactive or immobilized one.

The use of baited traps for studying blood preferences has long been practised in mosquito studies in general, and the most critical of these investigations have revealed many snags in interpreting the results. One of the first disclosures by workers in Malaya was that with some culicines attracted to such traps (man-baited) only a small proportion, less than 1%, actually fed on the human bait. At the same time other species attracted in much smaller numbers all fed readily. Parallel experiments from other parts of the world also disclosed the importance of trap design in host selection experiments. Even with a standard choice of host (e.g. man versus calf) one design may heavily bias results towards one host or the other.

Early studies on the culicine vectors of Japanese B Encephalitis in Japan were also concerned with the host role of birds versus mammals by the local vector *Culex tritaeniorhynchus*. In experiments involving seven different bird genera as compared to man, pig and other mammals, the influence of mosquito density was also disclosed. The heron-baited trap which was very satisfactory at high vector densities revealed certain limitations at low end-of-season densities, at which point the pig-baited trap became a more sensitive indicator of *C. taeniorhynchus* populations. These findings also endorsed those of others in showing that the engorgement rate of mosquitoes attracted to the bait can be quite independent of the attraction to that particular bait (Muirhead-Thomson, 1968).

Many of these snags in interpreting such host selection experiments based on baited traps have been revealed by the critical use of the precipitin test. Early work in Malaya showed that some baited traps were entered by mosquitoes which had fed on another type of host elsewhere. Clearly, one cannot assume, without precipitin test verification, that all engorged mosquitoes collected in baited traps have fed on the particular bait host. This is well illustrated by experiments in Pakistan on the attraction of *Anopheles culicifacies*, the main malaria vector, to human bait in a standard baited net trap (page 64), and to human bait in the open (Akiyama, 1973). In five bait nights no females were taken on the human bait in the open, but over the same period 78 females containing blood were taken in the human-baited net trap. When precipitin-tested, however, all were positive for bovid and negative for human blood.

The lesson from all this, and particularly from the recent researches in California and Florida, is that the very complex question of bait selection merits — and in fact demands — a multivalent approach, using all possible techniques. Ideally, these should be carried out synchronously in the same area, and with full awareness of the limitations, both known and unpredictable, and the advantages of each study method.

vi. . Combinations of Different Study Techniques

One of the most interesting recent demonstrations of the use of all available techniques to investigate an unusually complicated blood-feeding pattern was carried out in the Zika Forest, near Entebbe, Uganda, from 1966 to 1971 (McCrae and co-workers, 1976). This 300 acre (1.2 km^2) forest had been the scene of very intensive studies on the diel biting cycles of a wide range of mosquitoes for several years previously in connection with virus transmission work. The 120-foot tower constructed in the middle of the forest, with platforms at different levels, had provided a mass of information over the years about biting activities and biting cycles of many different kinds of mosquito, as well as of other biting insects such as Tabanids.

Among the mosquitoes included in these routine platform and ground level man-baited catches was *Anopheles implexus*, noteworthy more for being the largest African anopheline rather than for its incrimination as a vector of any significance. Most of that previous work had made use of human bait at different platform levels, but after many years it still had to be admitted that, apart from its regular recording as a man-biter in the forest, little was yet known about the true host preference of this species. For this and other reasons, *Anopheles implexus* seemed to provide a good subject for investigation, not only on the question of host preference but also as to all aspects of flight behaviour and host seeking. The course of the investigation is instructive in the way in which each particular problem was tackled in a series of phases, each of which logically followed on the other.

As previous records using the Zika tower had shown that biting by *A. implexus* in the forest is almost exclusively confined to ground level, the first stage of the study was to extend observations horizontally, by means of man-baited 24-hour catches carried out in seven types of capture sites representing a range of conditions within the forest and up to the forest edge. The most striking feature of these catches, carried out over a long period, was that well within the forest about 66-69% of the *implexus* were taken by day, whereas at the extreme edge of the forest this proportion fell to only 3% (Figure 4.4). A thorough recording of all relevant meteorological factors, such as temperature, relative humidity and light, showed that none of these factors, singly or in combination, could account for this marked disparity in feeding pattern.

A further series of tests were carried out in three sites typical of forest edge, 10 m within the forest, and in the centre of the forest, in order to allow direct numerical comparison between these sites. In this series no *implexus* at all were taken in daylight hours at the forest edge (Figure 4.5). Daytime biting activity at the centre of the forest was similar in general pattern to nocturnal activities at the forest edge, while at the site 10 m within the forest a more irregular pattern of biting activity by day and by night was revealed.

In view of the known influence of moon light phases on mosquito activity in general, catches were then carried out in accordance with different moon phases, and it was found that at each site the catches in the first half of the night were similar both in numbers and in trend at different moon phases, but in the second half of the night they diverged markedly, the numbers at full moon being much greater than at no moon, and the period of activity being considerable more prolonged.

The next phase of activity was concerned with the composition of the resting population. Fortunately, resting females could be taken readily in large numbers on tree trunks within the forest (up to 8-9 per man hour), and of these the unusually high proportion of 20% were found to contain blood. This provided an ideal opportunity for a critical examination of blood meal source, using both the precipitin test and the haemoglobin agglutination inhibition test (HI). Of the 1243 blood meals tested, over 96% gave positive results, and of these positives over 90% were bovids. There were only a few primate blood samples — probably from man — an no evidence that *implexus* had fed on birds, reptiles or amphibians.

The extremely high bovid proportion of blood presented a real problem in that there seemed to be no way of reconciling this with the scarcity of bovids within the forest. Cattle were only rarely and briefly brought into the forest for watering for about 10 minutes around noon each day, but no *implexus* were seen biting them on these rare visits. The other possible source of bovid blood, the Sitatunga antelope, was only present in very low numbers. By means of more specific tests on blood meals, enabling the blood of cattle to be distinguished from that of either Sitatunga or Duiker, it was confirmed absolutely that cattle and cattle alone formed the major source of blood for *Anopheles implexus*. Moreover, the only possible source of this

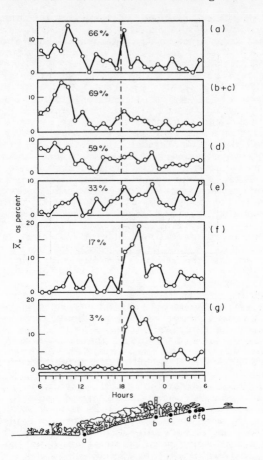

Fig. 4.4 Comparative daily patterns of *A. implexus* hourly incidence from all man-baited 24-hours catches conducted at points a-g across the Zika Forest as shown in the sectional view below. (After McCrae and co-workers, 1976).

cattle blood had to be outside the forest. *A. implexus* could therefore no longer be regarded as an exclusively forest dweller and, furthermore, it now appeared that the man-baited catches within the forest had not only employed a relatively unattractive host but had also been conducted in places and at times when *A. implexus* was seldom biting at all.

In order to carry the enquiry a stage further, a second phase of field studies was carried out. Daytime collections on cattle grazing close to the Zika forest edge showed that, as with man bait, biting was negligible; nor were resting *implexus* found in their vicinity in day searches. All-night collections were then started on groups of cattle at two grassland sites 150 m and 200 m from the forest edge. To check any possible bias of attraction to the team of collectors, blood meals of freshly fed females were precipitin tested and all showed cattle blood only. In these all-night catches away

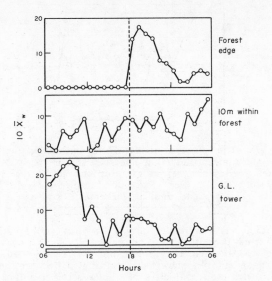

Fig. 4.5 Daily patterns of hourly incidence of *A. implexus*, from ten man-baited 24-hour catches at three locations, viz. forest edge, 10 m within forest, and centre of forest (G.L., tower). (After McCrae and co-workers, 1976).

from the forest edge, the biting rate of all mosquito species was on average 12.6 per man hour, that for *implexus* specifically being 1.5. Nearly twice as many *implexus* were taken at the nearer site (150 m) than at the further one (200 m).

On the cattle bait the mean nightly activity pattern was similar in several respects to that on man bait at the forest edge previously determined. As further confirmation that *implexus* seldom attacks man away from the forest edge, 24-hour catches using human bait were carried out at a similar distance from the forest edge as the cattle sites above. No *implexus* were taken.

The next step in investigation was to bring cattle as bait into the forest and carry out 24-hour collections. As mentioned above, this was an artificial situation in that cattle are normally brought into the forest for watering only, and for a very brief period round midday. On intervening dates, 24-hour man-baited collections were carried out in order to obtain man-biting data over the same calendar period. The cattle-biting pattern in this forest site was strikingly different from that on man already recorded in that two compact biting peaks were revealed, one in the three hours following sunset and the other in the hour following sunrise (Figure 4.6). The man-biting pattern showed the familiar indistinct rise in the first half of the day.

Having established the biting rhythm and host preference of *A. implexus* under a range of conditions inside and outside the forest, attention was now turned to a closer examination of the flight activity involved in this nightly exodus from the forest to seek a blood meal. Making full use of all the information which had been gained by that time regarding resting sites of *A. implexus* on tree trunks within the forest, hourly catches were designed to show rates of arrival at these favoured sites. On a number of suitable

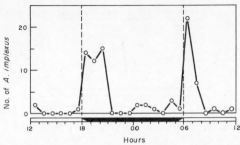

Fig. 4.6   The pattern of *A. implexus* hourly incidence
on cattle inside the Zika Forest.   From three 24-hour
catches each from a single bullock at sites at or
equivalent to b (see foot of Figure 4.4).   (After
McCrae and co-workers, 1976).

trees, all mosquitoes were collected hourly so that subsequent hourly catches
on the identical sites provided data on new arrivals.   Females collected were
graded according to whether they were unfed, freshly fed, half gravid or
fully gravid.   These figures showed a striking difference between night and
daytime numbers, with the majority of the resting population being absent
from the forest at night.   The numbers of resters often remained high until
late afternoon, but this is evidently followed by a massive dusk exodus from
the forest.   Arrival patterns also showed a massive post-dawn return from
outside the forest.

    The records also revealed considerable movement of resters, including
unfed females, between resting sites inside the forest during the day.   A
further study of this movement was made by using teams of collectors on
short shifts, moving to new sites at regular intervals.   From the moment each
team took up a new collecting position, all catches were segregated into 10-
minute periods and the results were used to determine the short-term pattern
of mosquito arrival at the man bait.   This casual biting on man within the
forest was evidently induced either by the arrival of the host man within the
attraction radius of resting unfed mosquitoes, or by the day-time movements
of such females bringing them into contact with the static bait.   This type
of biting activity was classed as passive or opportunistic host selection,
characteristic of behaviour in the presence of non-specific hosts such as man
or other animals within the forest.   In contrast to this is the specific
attraction to cattle, strong enough to stimulate the nightly migration from
the forest in search of preferred blood meals outside the forest.

    This thorough investigation, which dealt with many other aspects of
mosquito behaviour in addition to blood feeding activities — such as nectar
feeding, age-grading, etc. — naturally could not furnish a satisfactory
explanation for all the phenomena encountered.   Some puzzling and anomalous
observations remain unexplained, as is so often the case the deeper one
probes.   Nevertheless, the general strategy of attacking this problem by
applying all available knowledge of mosquito behaviour, and combining the
whole range of proved study methods, old and new, is obviously a very
profitable one which could be applied with advantage elsewhere.   One other
noteworthy feature of this investigation was that all experience and resources
were brought to bear on the special behaviour patterns of a single species of
mosquito, and with reference to only two hosts, viz. man and cattle.   Several
other studies on tropical mosquitoes described in this review have attempted
the difficult task of including the entire local mosquito fauna within one
compass, and have demonstrated that by extending the breadth and scope
there has almost inevitably been some loss in penetration in depth.

vii.   Special Host Selection Problems in the *Anopheles gambiae* Complex

The blood-feeding preferences of *Anopheles gambiae*, the main vector of malaria over most of the African continent, have been the subject of special attention for many years.   It was early realised that two brackish-water forms of *A. gambiae*, one on the West coast and one on the East coast, differed in several ways from the fresh-water form.   These two have now been accorded specific rank as *Anopheles melas* and *A. merus*, respectively. However, there were many pointers to the likelihood that the fresh-water form itself was not uniform, and could comprise two or more forms differing in certain behaviour features, in their degree of association with man, and consequently in their relative role in malaria transmission.   However, no consistent morphological differences could be found to distinguish *A. gambiae* from different areas in which such behavioural differences were reported, and accordingly little progress could be made towards a clearer understanding of the suspect 'biological races' involved.   In the last 15 years the situation has changed, and considerable advances have been made towards resolving this problem (White, 1974; Highton and co-workers, 1979).   Initially this was stimulated by the finding that *Anopheles gambiae* from different areas revealed the existence of 'strains' differing in their degree of susceptibility or resistance to the insecticide DDT, recently introduced for control of this species.   This finding, together with refinements in the technique of inducing mating in laboratory colonies of mosquitoes, launched a wide programme of cross-mating experiments with strains of *A. gambiae* from different areas in order to establish their specific identity as judged by the sterility or otherwise of the progeny of such matings.   By this means a broad definition of two main 'sibling species' could be made, and these were called provisionally Gambiae A and Gambiae B.   Further clarification was achieved by the finding that the chromosome structure in cells of the larval salivary glands differed distinctly in these two groups, and enabled a third species, C, to be defined.   Finally, it was found that the chromosome structure in certain organs of the adult female mosquito allowed these three main groups to be separated in the adult stage.

In most areas the bulk of the *Anopheles gambiae* was made up of varying proportions of A and B, while form C was more restricted in distribution to southern Africa and to fringe areas in the nothern parts of the distribution of the 'gambiae complex'.   A fourth form, D, has an even more restricted distribution in Central Africa.   The three main forms or 'cytotypes' of *Anopheles gambiae* have now been given specific rank as *A. gambiae* s.s. (A), *A. arabiensis* (B) and *A. quadriannulatus* (C).   For many years, however, these forms were studied as A, B and C, and it is these studies, with particular reference to blood-feeding patterns, which are relevant to the present discussion.

In the application of the precipitin test to the special situation of the *Anopheles gambiae* complex, there are certain preliminary steps which have to be taken which render the whole process much more tedious and less straightforward than any of the precipitin test records described above.   The particular banding pattern of the chromosomes which differentiates the females of the three main groups is only discernible in the special 'giant' or polytene chromosomes which occur at a particular stage in the development of the ovarian nurse cells, an integral part of the ovarian system (Figure 4.7). The course of development of these nurse cells is determined by the development of the ovaries themselves, which in turn is dependent on the digestion of the blood meal of the female mosquito.   In females with freshly engorged blood, the ovaries and the nurse cells are at too early a stage for chromosome identification;   consequently, there would be no means of indentifiying individual females whose freshly engorged blood had been expressed onto filter paper for precipitin testing.   In practice, captured gorged females have to be retained alive until the ovaries have developed to the precise

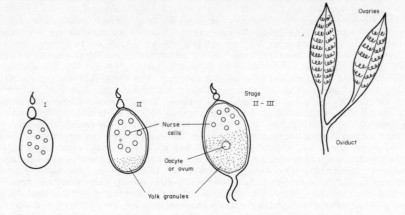

Fig. 4.7 Ovaries of female mosquito, and different stages (after
Christophers) in development of ovarian follicles.

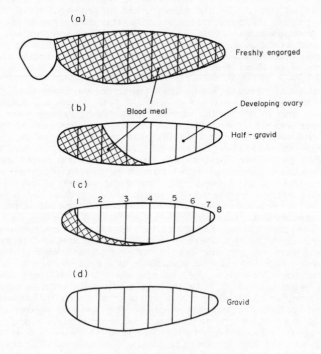

Fig. 4.8 External lateral view of abdomen of female mosquito showing
visible changed following the digestion of the blood meal and the
development of the ovaries.

stage when chromosome characters can best be seen in the nurse cells. This stage — which according to the conventional classification is Christophers stage III or early stage IV — corresponds to the 'half-gravid' phase in the gonotrophic cycle (pages 10–11) and can be judged approximately by external examination of the abdomen to see what proportion of it is occupied by the dark digesting blood meal, and what proportion by the paler develping ovaries (Figure 4.8).

Collections of live females from various shelters indoors and outdoors usually include individuals of two or three different stages in the gonotrophic cycle. Of these, the freshly engorged females are retained alive as described above. The half-gravids may contain some which are suitable for immediate dissection and identification, and their stomachs still contain blood in a fit state for subsequent precipitin testing of the extracted blood meal. Females beyond this stage and approaching the fully-gravid stage can no longer be identified, and of course they no longer contain an identifiable blood meal. Under these conditions the whole process of building up a valid precipitin test record becomes very arduous and time-consuming, as well as demanding considerable skill and experience in distinguishing the all-important banding pattern characteristics of the polytene chromosomes. In addition, not all the females examined at the critical stage reveal clearly identifiable chromosome banding.

The bulk of the precipitin testing of blood meals has been carried out on species A (*A. gambiae* s.s.) and B (*A. arabiensis*) in localities where their distribution coincides or overlaps. Most of these studies agree in showing that *A. gambiae* is much more anthropophilic than *A. arabiensis*. This is examplified by recent studies in the Kano Plain of Kenya which showed that 98% of *A. gambiae* (A) fed on man as compared with only 39% in the case of *A. arabiensis* (B). In the case of cattle, 59% of *A. arabiensis* recorded blood of this host, as compared with only 5% in the case of *A. gambiae* s.s. (Highton and co-workers, 1979).

The problem of defining blood feeding preferences of the different members of the *Anopheles gambiae* complex was also tackled experimentally in the earlier stages of the differentiation of this group. These experiments were designed and carried out in the East African coastal region with the object of comparing the reactions of insectary reared colonies of *A. merus*, the east coast salt-water member of the complex, and inland freshwater strains, with reference to their zoophily or degree of attraction to animals (Gillies, 1967). Different designs of experiment were carried out in a large room screened inside with plastic mosquito gauze and fitted with a cotton ceiling at the height of 7 feet. In the first series of experiments a stall for a calf was erected inside the screened area, and a bed was set up just outside the stall. In the second series, the room was divided by screening into three comparments. In one of the outer compartments a calf was stalled, while the other contained a bed for the sleeper. The narrow central section was used for release of mosquitoes, 50–100 at a time. The mosquito gauze walls dividing the releasing section from the outer compartments extended to within 4 inches of the ceiling, the gap allowing free access of mosquitoes into either of the outer sections. These gaps were closed just prior to the collection of mosquitoes in the morning. The two hosts were exposed over-night, and engorged mosquitoes collected in the morning had the stomach contents expressed onto filter paper for subsequent precipitin testing. When paired releases of different strains were made, these were marked with different coloured 'gold' and 'silver' dusts.

With this experimental set-up there were three stages in assessing different host preference reactions on the part of *A. merus* and an inland freshwater strain — which proved subsequently to be mainly species A. Firstly, when mosquitoes were allowed access to one host only, with no choice, they fed readily, but the proportion engorged differed according to the strain and the host. By itself, this preliminary test with no option of host did not give conclusive results, apart from showing that the hungry mosquito of either strain will feed readily on either host if that is the only blood source available.

The second series of experiments, i.e. with both hosts confined to the same compartment, and the third series in which they were in separate compartments, agreed in showing a significantly greater proportion of *A. merus* feeding on cattle than in the case of *A. gambiae*. However, although the differences in host preference were significant, with roughly 49% of *A. gambiae* A feeding on man as compared with 28% for *A. merus*, the reactions were not sufficiently marked to demonstrate convincingly either a dominant anthropophily of one species or a dominant zoophily of the other.

The experiments did reveal a very important variable factor in this type of host choice — namely, the variability of the human bait. One of the two human bait used had a much higher attraction to mosquitoes than the other — or was possibly a more passive and accessible sleeper — and consequently experiments involving the different human hosts produced widely different results. In order to deal with this variable, the mosquito marking experiment allowed two strains of mosquito to be exposed to choice of hosts simultaneously, so that whatever variables were implicit in the test itself, both strains were equally exposed and affected. The results corroborated previous findings, with 60% of *A. gambiae* A feeding on man as compared with 40% for *A. merus*.

In the experiments so far mosquitoes released from the central chamber had free access both ways to the other two chambers at all times. In one series of tests, the connection between the release chamber and the other two was provided with baffles, allowing mosquitoes to fly into the host chambers of their choice, but not to return. Under these conditions the proportions of *A. merus* and *A. gambiae* A feeding on man became approximately equal. It was clear that the introduction of the one-way baffles had masked the existence of the host preferences revealed by the previous tests. The interpretation of this was that when free flight was unimpeded, random flight initially took many *A. merus* into the man-host chamber. But many of those mosquitoes then left that chamber without feeding and moved into the animal chamber without hindrance, where feeding took place more readily. If those mosquitoes are prevented from leaving the man chamber after this initial random flight, a high proportion will eventually feed on man as the only available host.

Precipitin testing carried out with regard to the final location of gorged females in the three compartments in the test room, in the absence of impeding baffles, also revealed that a small proportion of *A. merus* (3–4%) which had fed on man in the man compartment later moved into the animal compartment; and an equally small proportion of females which had fed on the calf were later found in the man compartment. In the case of the *A. gambiae* A, rather more mobility appeared; of the total feeding on man in the man compartments 83% remained in the man room, while 11% moved to the animal compartment, 6% returning to the central release compartment.

The main object of these experiments was to devise some kind of standard behavioural or host-preference test which could provide a clear, consistent and measurable difference in behaviour patterns between members of the *Anopheles gambiae* complex. It appears that in the design of such tests results may be greatly affected by such seemingly small technical details as presence or absence of baffles in the communication between release and host chambers. The relative attraction of different human hosts employed is also a major, and less controllable, factor. Also important is the amount of space separating the animal and human host, results indicating, for example, that when the human sleeps just outside the calf stall, his particular odours are overwhelmed by those of the animal. It is possible that other variables, overlooked or unsuspected, were also playing a part in masking or in exaggerating behavioural responses.

Unfortunately, this promising line of controllable experiment involving female mosquitoes of known age, condition and strain, has received little close attention with regard to the *A. gambiae* complex since that work. It would still seem to have an important part to play in unravelling behaviour patterns of closely related members of this and other complexes. As far as blood feeding habits are concerned, future experiments will probably have to aim at more uniformity in the nature of both the human and the animal host provided, possibly by relying less on results shown with single hosts — with their personal, and variable, degree of attraction — and use instead groups of hosts in order to minimise the upsetting effects of individual variability.

viii. Host Selection Within the Human Family Group

In most studies involving the use of human bait to attract mosquitoes, and biting flies in general, it has been the experience that individuals tend to differ in their attraction. Some individuals are consistently more attractive to mosquitoes than others, while in some cases the attraction of a particular individual, relative to others, may fluctuate from day to day (Freyvogel, 1961; Khan and co-workers, 1971). Where man-biting rates are used as a measure of the biting population, or of the population in general, the existence of these individual variations is acknowledged, and in most cases counteracted by the use of more than one human bait at a time. In endeavouring to understand or explain these individual differences many factors have had to be considered, and all lines of enquiry eventually lead back to the basic problem of what attracts mosquitoes to human hosts in the first instance (Freyvogel, 1961; Gillett, 1979). This basic problem in turn is one affecting the feeding responses of all man-biting flies and blood-sucking insects and constitutes by itself a vast and proliferating subject, impossible to so full justice to within a book of this nature. In addition it is a subject where laboratory experiments with controlled environments and controllable individual physical and chemical factors play a vital role; this again puts much of the work beyond the scope of the present review, in which the emphasis is on field study methods rather than laboratory ones.

One aspect of this differential attraction of the human host is particularly suited for closer examination as it is based on observations and experiments in the natural environment of humans living in rural conditions in hot countries and exposed in the normal course of their lives to a high degree of mosquito biting. This aspect also is an instructive example of the development of progressively more sophisticated techniques to bear on essentially the same problem.

Until about 1950 the bulk of observations on the differential attraction of human subjects was based on adult or adolescents (bait boys) or older children. Nothing was known about the existence or otherwise of the mosquito's preference for particular age groups of humans. For many years malariologists and epidemiologists had to assume that, apart from the comparatively minor individual variations, mosquito biting among humans was a more or less random process. It was only comparatively late in that period that some doubts were cast on this assumption due to certain discrepancies between theoretical calculations regarding the extent to which exposure to the bite of infected mosquitoes increased with the age of the human subject, and actual parasitological studies on human populations in areas of high malaria transmission.

Opportunities for investigating this problem by direct observation in the field first arose in Jamaica in 1950 due to a combination of favourable circumstances (Muirhead-Thomson, 1951). Firstly, the main mosquito vector of malaria in that island, *Anopheles albimanus*, is present at very high adult populations at certain periods of the year, with a high biting rate on man. Secondly, while this mosquito will bite throughout the night there is a very sharp peak of biting, lasting for 20-30 minutes, at sundown, when there is still sufficient light to see mosquitoes landing on or biting exposed human subjects sitting outside their houses. The third favourable factor was the presence of rural family groups, of both negro and east Indian origin, who cooperated in these experiments. The results of the first series of observations on a whole family sitting outdoors at dusk are shown in Table 4.4(*), and show that many bites were received by all members of the family except for zero biting on the one-year old infant. Also striking is the dominant attraction of the adult male householder over his wife and the rest of the family. These observations were repeated under conditions which demanded an even closer cooperation on the part of the householder, namely collections of mosquitoes biting the sleeping members of the family indoors during the half hour before dawn — the six children being asleep on the same bed, with the father sitting nearby, the wife being absent. The results, in Table 4.4(**), again show the dominent attraction of the adult male. With regard to the baby Linette, observations showed that the few mosquitoes which did attempt to bite spent a much longer time probing around than in the case of the older children. Even when asleep the sensitivity of the baby, previously noted outdoors at dusk, was strongly marked, and the odd mosquito attempting to bite was often shaken off without the infant waking.

Table 4.4 Distribution of *Anopheles albimanus* bites among a
Jamaican Negro family:
* sitting outdoors during the 20-30 minute peak of biting at dusk;
** sleeping indoors at night in the pre-dawn period.
(After Muirhead-Thomson, 1951)

Subject	Sex	Age	Summer weather* (5 collections)	Rainy season* (1 collection)	Rainy season** (1 collection)
Mr Bailey	M	36	103	199	81
Mrs Bailey	F	30	25	49	–
Marjorie	F	9	23	40	3
Donald	M	8	45	61	13
George	M	5½	8	13	0
Thelma	F	5½	18	15	1
Malcolm	M	2½	18	15	1
Linette	F	1	0	2	3

Similar observations carried out on families of East Indian origin in Jamaica, and in Trinidad where the vector species involved were *Anopheles bellator* and *A. aquasalis*, showed that the low to zero attraction of infants was marked and consistent in the approximate age group of 12 months and under. From the age of two years onwards biting on children of different age groups was much more frequent and did not fall into any consistent patterns. Younger children often received more bites than older children in the same family group under equal conditions of exposure. Particular interest was attached to the mother-infant combination of hosts, as these two spend so much of the day and night in close association. With the seven different family groups observed in these two islands, in each case the mothers attracted a large number of mosquito bites compared to the low to zero biting on their babies.
Since that time similar types of study have been carried out on anophelines in other countries, particularly in Africa (Carnevale and co-workers,

1978) and in Papua New Guinea (Spencer, 1967), using the same technique of catching mosquitoes attempting to bite individuals of different age groups exposed together. In extensive studies near Brazzaville a marked increase of attraction with age of human host was observed with *Anopheles gambiae*, with adults three times more attractive than infants. In the Papua New Guinea studies collections were made on eight mother and infant pairs sleeping side by side, and it was found that none of the infants were bitten even though the mothers received over 100 bites during the study.

In host selection experiments of this kind, the presence of the human observer/collector is unavoidable, but nevertheless it presents a factor whose influence is difficult to gauge and cannot be ignored. Bearing in mind this need to eliminate the human presence, several attempts have been made to use blood grouping techniques as a means of identifying the source of the mosquito's blood meal. The most recent and successful refinement of this technique is based on a particular group of serum proteins called the hapto-globins (Hp) (Boreham and co-workers, 1978; Port and co-workers, 1980). Many people lack this Hp factor, and where it is present such bloods can be further subdivided into four groups by serological methods. In addition, the more conventional AB and O blood groups can be identified in mosquito blood meals. Both of these methods can be used to study selective feeding on different human hosts in the absence of a human collector, and both were used in an intensive study carried out in the Gambia in 1978 (Port and co-workers, 1980). The first step in this investigation was to carry out a blood census of the 150 people living in the African village selected, with reference to the ABO and Hp types in each hut or habitation. When these types from two or more individuals differed so that it was possible to identify each of them, these individuals were used to form experimental groups. Twenty groups were finally selected, eight separated by Hp, four by ABO and eight by both methods. As far as possible these groups were chosen from individuals who normally slept in the same room or under the same roof. The groups were mainly mothers and children, although combinations of children, adolescents and adults were also used. Each group was asked to sleep under mosquito nets of a type which were not adequate to exclude all biting mosquitoes but which acted as a trap for those mosquitoes which did penetrate. The mosquito net was let down in the morning and gorged mosquitoes collected from inside.

A total of 2339 blood meals were typed during this study, and 92% of these were positively identified. The experiment not only confirmed previous observations on *Anopheles gambiae* in Kenya that mothers are fed on more frequently than their babies, but also provided a great deal of information about the relation of host size to mosquito feeding, and the extent of multiple feeds in which the same mosquito had fed on more than one of the available hosts during the same night. While those figures conform to the same general pattern experienced elsewhere as to the much greater attraction of mother over infant, the infants in the African experiments do appear to be bitten to a much greater extent than those in either the West Indian experiments or the Papua New Guinea series. This may well be a reflection on the more voracious man-biting habits of *Anopheles gambiae*, not only with regard to biting infants but also in obtaining a blood meal from them.

Observations on other groups of mosquitoes in Britain have shown that specific differences exist with regard to the three main phases of feeding on a host; namely, an exploratory period after the mosquito lands, a pene-tration period, and finally a feeding period which begins with the actual appearance of ingested blood in the mouthparts (Service, 1971b). The explor-atory period may only last 8–11 seconds with some species, but considerably longer with others. The penetration period in several species is less than a minute, but is much longer with others. Marked differences also exist between species with regard to the actual feeding period: 82–87 seconds for some *Aedes* species, but longer in other mosquitoes observed.

These and other behavioural differences might determine the extent to which the relatively few mosquitoes which attempt to bite the sensitive infants do or do not succeeed in obtaining a blood meal. Clearly, the new serological techniques have provided some of the answers to this fascinating problem of host selection within the family group. However, there would still seem to be a need for direct observation on mosquito reactions on approaching these different human subjects, and there is no doubt that this could be achieved without introducing the experimentally undesirable human odour of the observer. Workers in the field of both tsetse and blackflies (Simulium) have found it necessary to devise special methods of nullifying the physical and chemical presence of the human observer in host-feeing experiments, and the success of their methods should encourage the use of similar study methods for biting mosquitoes in their natural environment.

CHAPTER 5

Tsetse Flies

i. Introduction

The tsetse flies of Africa have been one of the most intensively studied groups of insect disease vectors.   As the transmitters of both human and animal trypanosomiasis of domestic stock they have played, and continue to play, a vital role in all movements, activities and settlements involving man and his cattle.   The capacity of many species of tsetse to feed on different game animals in Africa — some of which are reservoirs of trypanosome infection — and the dominant role played by vegetation types and associations in determining distributions and movements of tsetse, all contribute to making the study of tsetse ecology and behaviour a peculiarly difficult one, and at the same time a fascinating one.   Because of the many facets and ramifications of the tsetse problem, it has attracted a long succession of competent specialists from   many   disciplines   = entomology,   parasitology,   zoology, ecology, medicine and public health, as well as geographers and all those concerned with rural development and settlement.

81

This vast field of study to which the British, French, Belgians and Portuguese have been major contributors, has produced many outstanding books which form essential reading for anyone being introduced into this field, as well as for those research workers who have only been able to concentrate on one particular aspects of the problem, or whose studies have unavoidably been restricted to certain species of tsetse or certain parts of that great continent (Buxton, 1955; Ford, 1971; Glasgow, 1963; Mulligan, 1970).

For present purposes, no attempt will be made to cover or even condense available knowledge about tsetse ecology and behaviour; the sheer bulk of many standard text books on the subject shows how futile this would be. The emphasis in this chapter will be once more not so much the knowledge that has been gained about tsetse but rather the methods which have been used to build up this mass of information. A few essential points will help to introduce this.

Tsetse flies are large robust insects, almost entirely diurnal in activities, and consequently easily visible, especially when settled. Unlike mosquitoes, midges and sandflies, both males and females are blood suckers. In the case of the female the blood feed is an essential part of the gonotrophic cycle in which the digestion of the blood meal leads to the development of the ovaries to maturity. This in turn is followed by the production, not of eggs, but of a late instar larva deposited on the ground, and which has only a brief larval life before pupating in the soil. With a further blood meal the cycle starts again.

In the case of the male, the digesting blood meal is stored as fat which is later used up in flight activity and blood seeking. In this process the male passes through a hunger cycle corresponding in some ways to the gonotrophic cycle of the female. The behaviour of tsetse is greatly influenced by the particular stage in these cycles, and this has to be taken into account when interpreting observations or captures in the field.

Although tsetse, because of their relatively large size and daytime activity, can be seen visually near by, or even at some distance with the aid of a telescope, direct observation is not always a feasible method of study. This is because of the fact that while some species give the appearance of being very abundant, swarming round large animals or vehicles passing through their woodland habitats, others normally exist at very low densities making it necessary for routine captures to be carried out over long rounds or transects.

As with many other groups of insect disease vectors, basic knowledge about the distribution and seasonal abundance of tsetse is based on a variety of capture or sampling methods, many of which have evolved independently according to the species being studies, the object of the research or survey, and the particular preference or ideas of each observer. In many cases the basic capture techniques developed have provided the first ideas about tsetse behaviour, in which capture data with regard to proportion of males and females, different hunger cycle stages, age grading, etc., are compared to other sets of figures obtained at different times or under different conditions in order to indicate daily or hourly movements of flies in relation to food, shelter and mating activities.

In the evolution of these sampling methods, tsetse workers have often followed courses quite independently of workers faced with similar problems in another part of Africa. Tsetse workers in general, often engaged in a lifetime study of this one group, have had little time to equate their techniques and experience to other fields of medical entomology. This in turn has often made it difficult for the newcomer to this specialised field to intrude into an area in which every possible aspect of tsetse biology and behaviour seems to have been covered thoroughly at one time or another by established authorities. However, the last ten years has witnessed an enormous upsurge of new ideas into this well-trodden ground, and has revolutionised many longstanding study methods and their interpretation in a manner which has had a profound effect on many other fields of insect vector study.

ii.   Review of Long-established Tsetse Capture Methods

Before proceeding to discuss these new developments in greater detail, it will be necessary to review briefly some of the longstanding field methods for studying tsetse ecology and behaviour, with special reference to four main procedures, viz. the fly-round, the use of stationary bait (animal), resting populations of tsetse, and the use of artificial or simulated animal traps. For further details of those earlier publications, the reader is referred to an earlier introduction and critical review of the study methods involved (Muirhead-Thomson, 1968).

The fly round.   The fly round has been standard procedure for many years, with special reference to the game tsetse of East Africa, particularly *Glossina morsitans* which ranges far and wide in savannah woodland and is one of the main vectors of animal trypanosomiasis affecting cattle. *G. morsitans* is also implicated in outbreaks of the virulent 'Rhodesian' form of human trypanosomiasis which flares up locally from time to time.

In the fly round a path is laid out crossing a range of vegetation types. A group of fly boys who act as combined bait and catchers move along the path stopping from time to time to collect the tsetse which follow them or are attracted to them. Alternatively, two collectors may carry a cloth screen between them, catching the flies on the screen and on themselves with hand nets. Normally the attractant is the human bait, but cow or ox can be used in the fly round in order to collect another important savannah species, *G. pallidipes*, which is less attracted to man only. This has been a widely used technique in tsetse fly surveys for many years, and is still in use in some parts of Africa both for surveys and for the evaluation of control measures. It has also been applied in the form of a linear 'round' for sampling the riverine species of West Africa, *G. palpalis* and *G. tachinoides*, which tend to be restricted to dense riverine belts of gallery forest and are also species closely involved in the transmission of human trypanosomiasis.

The original fly round has been modified from time to time in order to meet special requirements. In the transect fly round developed by some workers in East Africa, the fly round follows arbitrary straight lines which are divided into numerous short sections of equal length. The catching party halts and collects only at the points defining the sections. In this way it was hoped to arrive at a more standardized technique, less subject to the vagaries and variables of the human collectors. The Belgian workers, also anxious to establish a more standardized fly round in order to obtain more representative samples of *Glossina morsitans* in the region that is now Zaire, laid out a plan involving four sides and two diagonals in a number of squares 100 x 100 m, each square forming part of an essembly of squares whose extent was determined by the variety and distribution of habitats.

Stationary bait (animal).   As an alternative to the fly round, some workers have preferred to use a stationary animal as bait, and to extend the collecting period up to eight hours or even the whole day. This method has several advantages in particular situations. The extension of the collecting period allows for any fluctuations in tsetse density during the day, and also enables periods of normal peak biting activity to be defined. In this respect it is particularly useful for species like *G. pallidipes* with a marked evening peak of biting. The method has the further advantage in that it catches female tsetse to a much greater extent than the fly-round, which produces predominantly male samples.   By allowing attracted flies of both sexes actually to feed on the ox before they are captured, the sample can be interpreted more accurately in terms of hungry flies (Pilson and Leggate, 1962a; Pilson and Pilson, 1967).

A logical follow-up of this technique is the extension of the catch to a full 12-hour diurnal period and to the complete 24-hour diel cycle.   This has enabled more accurate comparisons to be made between the daily biting cycles of the main vector species.   *Glossina pallidipes*, for example, shows a steady increase in biting activities throughout the course of the day, with a peak in the later afternoon;   *G. palpalis* has low activity in the early morning and late afternoon, with peak biting activity in the middle of the day; while *G. brevipalpis* has very low biting activity throughout the day but exhibits two sharp peaks at dawn and at dusk (Harley, 1965).

Trapping.   A variety of traps have long been used in tsetse studies, many of them taking their names from their originators — the Harris trap, the Chorley trap, etc.   One of these designs, the Morris trap, was the subject of a particularly penetrating series of tests and trials after it was first designed in West Africa over 30 years ago.   Much of that work has particular relevance to the general problem of sampling tsetse flies, which has been the subject of so much critical reappraisal in the past 10-15 years (Bursell, 1973;   Ford and co-workers, 1972;   Muirhead-Thomson, 1968).

The Morris trap (or 'animal' trap, as it was called by the originator) was originally designed to resemble a small host animal about the size of a goat (Figure 5.1).   The most important feature was the cylindrical shape of the body, producing highlights and shadows.   The curved body of hessian or cloth was open at the bottom.   Flies attracted into this opening from the undersurface of the body were directed to a slit along the upper surface of the body, over which a cage was superimposed.   A non-return device trapped

Fig. 5.1   The standard-size 'Morris' trap used for sampling tsetse flies. (After Morris, 1963).

all tsetse entering the cage. When these traps were sited carefully, with due regard to visibility from a wide angle and presence of natural hosts in the locality, they proved to be very productive. Comparative trials in West Africa and in East Africa — where a slightly larger 'animal' was found necessary to attract the species there — showed that for most of the year animal traps were much more productive than fly rounds and fly boy catches being carried out at the same time. Even more important was the fact that female tsetse made up a higher proportion (up to 82%) of the Morris trap catch.

The two sexes of tsetse are known to emerge from puparia in approximately equal numbers, but as females have a considerably longer life span than males a wild population of tsetse is reckoned to contain 70-80% of females. With all the species of tsetse tested the proportion of females taken in the Morris trap is very close to that estimated in the natural state, and this would appear to indicate that the trap sample is a good representation of the whole population. In contrast, one of the peculiar features of the fly round with its human bait is that for savannah species of tsetse, males make up the bulk of the catch, perhaps up to 90%. From the nature of the fly round one might assume that the basic principle is the attraction of hungry flies to bait. However, the situation appears to be rather more complicated than that. Many of the males caught in the fly round are not hungry, and it has therefore been assumed that they follow the bait or moving object as part of a behaviour pattern in the search for females. This peculiar sampling bias in the fly round has long been recognized and acknowledged by tsetse workers (Bursell, 1970, 1973; Buxton, 1955; Ford and co-workers, 1972). Despite that, a great deal was read into these figures from fly round catches; a convenient standard was adopted in the form of 'the number of non-teneral males collected per 10,000 yards' (the 'teneral' or recently emerged, soft-bodied males being discarded), and this figure was referred to as the 'apparent density'. These obviously male-biased samples of tsetse from fly rounds also formed the basis of extensive mark-release-recapture experiments in the 'thirties, the figures resulting from these being used in turn to provide longevity data and to calculate the expectation of life for tsetse.

To return to the Morris trap, it appears that it catches much more representative samples of the natural tsetse population, and might therefore be expected to be a more accurate pointer to determining normal behaviour patterns of different species. Again, nothing in tsetse work is quite as simple as it seems at first. When the Morris 'animal' trap was compared with live cattle bait and with the fly boy round in a three-pronged trial in East Africa, it was concluded that the traps were also sampling a different proportion of the population from that attracted to the cattle. For example, the time of day when most flies were taken showed differences between traps and cattle; with the cattle most flies were taken in the morning and evening, while in the traps most were taken between 12.20 and 14.00, with least in the morning. Further confirmation that different populations were being sampled was made possible with the new age-grading technique whereby females could be graded up to the fifth cycle of ovulation, i.e. up to the time when the insect is about 50 days old. When applied to trials employing four different sampling techniques, it was found that female G. pallidipes entering Morris traps were older than those caught in the fly round, those caught on bait animals being intermediate in age structure. Correlated with age structure was an increase in the proportion of females carrying third instar larvae. In general, therefore, the oldest females and those most advanced in pregnancy were more readily taken in traps, while the youngest and least advanced in pregnancy were more liable to be taken in the fly round.

The Morris traps, despite their animal-like appearance and design, were not simply functioning as substitute animals; they must also have been providing additional attraction to those flies, especially females, seeking a shaded resting place in the hottest brighter period of the day.   Later work with these traps in Liberia further confirmed that the theoretical basis for the attraction of such an 'animal' trap is still very doubtful.   It was observed that although the trap was essentially designed to resemble a host animal, in that particular study area the live animal itself (viz. goat) was not attractive to *Glossina palpalis*.

This briefly summarizes the state of progress by the early 'sixties with regard to advances in the understanding of some of the basic field methods in use for establishing behaviour patterns of different species of tsetse. Considerable progress had also been made in defining the normal day and night time resting sites of tsetse (Pilson and Leggate, 1962b) and in analysing these populations relative to those already discussed, all of which are concerned with the active or flying population.   The resting aspect of tsetse behaviour will be discussed later in the context of subsequent progress.

The main reason for laying greatest emphasis on the three main methods of sampling the active tsetse population — viz. fly round, stationary ox bait, and simulated 'animal' trap — has been that an understanding of this background is essential for a full appreciation of the course of events from the late 'sixtires until the present.

Progress in tsetse research is a continual process on a broad front, and it would be impossible to do full justice to all aspects of progress in the last 10-15 years.   On the subject of ecology and behaviour alone, publications have proliferated to such an extent that some rigid selection is unavoidable if one aims at presenting this material in a manner comprehensible to those not actively involved in tsetse research.

iii.   The Rhodesian Work on New Study Methods

There is little doubt that the most significant advances bearing on the theme of this book are those which have taken place in Rhodesia, where research workers have re-examined old problems in an entirely new and critical light.   Rhodesia has long been a great centre for tsetse research, and to this was added a further stimulus and incentive when several highly experienced tsetse workers from East Africa moved down there when those countries gained their independence.   Those strong teams in the Tsetse and Trypanosomiasis Control Branch of the Rhodesian Department of Veterinary Services were supplemented by a tsetse-oriented Zoology Department in the University of Rhodesia which had an international reputation in the field of tsetse physiology.   All this provided an ideal springboard for a new school of younger ecologists and entomologists to contribute new critical ideas and standards.   On top of this is the undoubted rallying spirit of scientists in a country ostracised politically, but fortunately not scientifically, from the rest of the world and determined to show that they could hold their own in adversity.

It would be useful to take as a starting point the stationary ox bait sampling method and the well-established fact that, in contrast to the fly round technique, both sexes were almost equally represented in the feeding fraction of the catch.   This technique was extensively used in Rhodesia in the early 'sixties for both *G. morsitans* and *G. pallidipes*, mainly with the object of investigating 'trypanosome challenge', i.e. the extent to which cattle were exposed to the bites of infected tsetse (Pilson and Pilson, 1967). In applying this technique to another area of Rhodesia in 1966, certain inconsistencies appeared in the results with regard to the sex ratio of *G. morsitans*, which might have been due to lack of skill or other human error in catching flies off bait (Phelps, 1968).   In order to check this

Fig. 5.2   The falling cage trap for tsetse flies as used in the
Rhodesian studies.   (After Phelps, 1968).

possibility a large net was used, 9 x 6 x 7 ft, which could be raised on a
pulley and then quickly lowered over a tethered ox, allowing one person to
remain inside to collect the tsetse (Figure 5.2).   The practice was to lower
the cage as soon as the animal arrived on site and thus capture flies
attracted to the moving bait.   The cage was left down for 10 minutes while
flies were collected and then raised for 10 minutes to expose the animal, and
then lowered again to allow the capture procedure to be repeated.   This
sampling at regular intervals continued throughout the day from 6 am to
6 pm, and in all months of the year.   Records were kept of total catch of
males and females, and the percentage of each sex which had gorged.
     Several facts emerged from this study.   The use of the falling cage to
ensure that all flies on the ox bait are caught still showed an unexpectedly
low proportion of gorged females in this experimental area in the Zambesi
valley near Lake Kariba.   Only on one occasion did females form 50% of the
total catch.   The proportion was usually well below this level, particularly
in the wet season when proportions down to 10%, and even to 0%, were more
usual.   It was clear that the possible factor of poor catching on the part of
the operator could be ruled out as an explanation of low female catch.   It
was obvious too that females were coming to standing ox bait for reasons
other than immediate feeding.
     The figures also revealed that the composition of the first falling net
catch of the day, made immediately at the end of the ox bait's walk to the
station site, differed from that recorded on the stationary animal for the
rest of the day in that a smaller proportion of females were caught but a
higher proportion of these had blood fed.   It was concluded that a moving
bait animal attracts hungry flies more readily than those which are not
hungry and that, as far as the stationary ox bait is concerned, females are
attracted for reasons other than immediate feeding.

From the operational point of view, some of the advantages of the falling cage technique as used for stationary bait are offset by the fact that while the net is lowered, access to the bait on the part of attracted flies is denied for the 10 minute collecting period. On the other hand, in the case of direct continuous catch the advantage of having the bait exposed to tsetse without interruption is possibly offset by the fact that at high tsetse densities even experienced catchers cannot be guaranteed to catch all the tsetse that alight.

The new queries raised as how best to interpret tsetse catches on ox bait, and the complicating factor of the composition of the catch being determined by whether the bait was moving or stationary, led naturally to the next phase of critical experiments from 1968 onwards.

In this series, the same basic problem of interpreting capture and sampling data in terms of tsetse behaviour patterns was approached from a slightly different angle (Vale, 1969). This originated from the well established observation that *Glossina morsitans* shows a feeding preference for certain game animals, particularly the warthog, despite the relative scarcity of this host in most game populations. It was decided to study these tsetse responses by means of artificial models of host animals, rather after the manner of the original Morris of 'animal' traps.

The basic technique was to use a 10 gallon black drum mounted horizontally on a small pram chassis about 14 inches above the ground. The model was moved along a straight 'run' at about 1.5 mph by means of a long handle held by an operator walking 23 ft ahead of the model, or alternatively by means of a long rope operated from a distance. The trials were carried out in runs representative of different types of woodland favoured by tsetse. The attractant for tsetse took various forms, either the model alone, model accompanied by walking man, or walking man alone. Tsetse visiting the attractant were counted approximately with the naked eye, or with the aid of a telescope, or by a sticky deposit previously applied to the attractant, or by the use of a large drop net operated as described above near the end of the run.

The first striking and consistent observation was that more tsetse, *Glossina morsitans*, were attracted to the unaccompanied model than to the walking man. Even more surprising was the observation that the model alone was much more attractive than the combination of model plus man walking three feet away (Table 5.1). In a series of carefully designed and replicated experiments, all possible combinations of attractant and tsetse capture methods were tried out, such as sticky deposits on the attractant, hand net catching and drop net. The results all pointed to certain conclusions of

Table 5.1   Total recovery of *Glossina morsitans* from thee daily replicates of experiment employing mobile attractants.   (After Vale, 1969).

	Male morsitans	Female morsitans
Recovery by falling cage		
Model alone	520	124
Model plus man	251	12
Man alone	110	4
Recovery by sticky deposit on back (4 experiments)		
Model alone	1168	944
Model plus man	733	151
Man alone	841	36

great significance to the whole problem of sampling tsetse.   For this species
of tsetse certainly, the presence of man close to the model attractant had a
depressing effect on the catch, particularly of the females.   Further observations showed that the effect on *G. pallidipes* was even more marked, affecting both sexes and especially the females.

In the next series of experiments a comparison was made between
several combinations of attractant: (i) a walking man who stopped at
frequent intervals to recover the attendant tsetse with a hand net;   (ii) a
walking man wearing a sticky screen on his back;   (iii) a man riding a
bicycle and stopping at intervals of 200 yards to catch attendant tsetse with
a hand net;   (iv) similar to (iii) except that recovery was effected by a
sticky screen bolted to the cycle frame so that it was close to the cyclist's
back;   (v) a Land-Rover pickup travelling at 10-15 mph: two continuously
vigilant catchers sat in the back to recover attendant tsetse with a hand
net;   and (vi) similar to (v) except that a 3 x 1 ft screen covered with
khaki sticky deposit and fixed to the back of the vehicle was used for
recovery instead of a hand net.

The idea of this layout was to compare the efficacy of hand-net captures of tsetse with that of sticky adhesive surfaces, under three different
conditions.   The results are shown in Table 5.2.

Table 5.2   Total recovery of *G. morsitans* using two different recovery
techniques for three different conditions.   (After Vale, 1969).

	Cyclist		Land Rover		Walking man	
	Males	Females	Males	Females	Males	Females
Hand net	120	9	684	69	123	6
Sticky surface	1307	63	1234	64	610	48

Under all these conditions it was clear that the recovery of tsetse by
sticky trapping was markedly superior to hand net catching.   In all these
experiments man was present in the role of attractant bait.

In view of the superiority of the sticky surface method of trapping
attracted tsetse, a series of experiments were then carried out in which the
black drum or 'model' previously used was treated with sticky deposit, and a
comparison was made between (i) this alone and (ii) a combination with a
man walking three feet away and wearing a sticky screen, (iii) a combination
with a man walking three feet away but not wearing a screen, and (iv) a
non-sticky drum accompanied by a man who caught with a hand net any
tsetse landing on the model.   These were all operated along traverses of
700-1400 yards passing through a variety of tsetse woodland.

In all these experiments the unaccompanied sticky model (i) trapped
many more males and females, and a higher proportion of females, than were
captured from the non-sticky model by hand net (iv).   Since the model plus
man combined, with sticky recovery, gave catches intermediate between those
from the sticky model alone and those from the model plus man using hand-
net recovery, it was concluded that the low recovery by the hand net technique was due both to the depressing effect of the catcher and to the
inefficiency of the hand net technique itself.

Much the same finding was made when a conventional fly round with
two vigilant catchers carrying a screen of black cloth was compared with the
catch on an unaccompanied sticky model over the same traverse.   In each
case the unaccompanied model trapped more males and females, and a much
higher proportion of females, than were taken in the fly round.

It is evident from these preliminary experiments that several new factors had been revealed which had a vital bearing on the validity of several conventional sampling methods. Consequently any concepts about tsetse behaviour patterns based on such data needed re-appraisal and re-examination. The important finding that with *G. morsitans*, and particularly with *G. pallidipes*, the presence of man — whether in the role of bait, catcher or observer — had a diminishing or even repellent effect on tsetse attracted to the model bait, helped to explain some previous inconsistencies. To explain the puzzling fact that the repellence is partial, and not absolute at any time, it is pointed out that other experiments indicate that man is probably not the normal host of *G. morsitans*, but that the non-attractance or even repellence to man may be overcome in the case of extremely hungry flies, mainly males, which have been unable to obtain a blood meal elsewhere.

In an attempt to analyse the depressing or repellent effect of man, life-size models of a man were made from tins of various sizes, painted in non-shiny black and clothed in a khaki uniform. They could be assembled upright, or arranged on hands and knees. Both live and model baits were fitted with sticky surfaces to trap alighting tsetse. The man model on hands and knees recorded much lower catches, indicating that odour must play a part in the repellent action. In the case of the upright stance, the model man showed the same degree of non-attraction as a live upright man, from which it was concluded that visual stimuli also play a part in repellence when the human is in the upright or vertical position.

iv.   This reaction to the presence of man on the part of tsetse in increasing order of intensity by *Glossina morsitans* males, *G. morsitans* females, *G. pallidipes* males and *G. pallidipes* females, dictated the course of further investigations. Man had to be excluded from any experimental set-up where his presence was likely to interfere with normal behaviour patterns, and accordingly the emphasis was now on the use of a live animal (ox) or a simulated animal 'model' as the standard attractant. With regard to capture of tsetse attracted to such bait, the emphasis was on the use of such purely mechanical methods as sticky surfaces in preference to any method depending on the presence of human catcher or observer (Vale, 1974b,c).

In further extensions of these experiments, the sticky compound used so far was itself critically examined to ensure that it did not possess some intrinsic attraction. In fact, at very step in this series of studies, every possible source of error or bias involved in the introduction of new techniques was critically tested so that the errors of interpretation in the past did not re-appear in a new and unsuspected guise.

In further studies on both *G. morsitans* and *G. pallidipes* in Rhodesia the original oil drum model was replaced by a horizontal cylinder of metal or fibreglass, 50 cm long and 37 cm in diameter, completely covered with cloth and mounted 37 cm above the gound on a metal frame, the frame being provided with spoked wheels when used for mobile tests (Figure 5.3). In addition to the use of sticky deposits, now coloured to match the bait, entirely new techniques were introduced based on the electrocution of flies making contact with electrified grids or surfaces. Run from 12-volt batteries, the electrified grids of blackened fine steel wire running 0.8 cm apart could be set up over the attractant, for example on a wooden board on a man's back or used in conjunction with fine nylon netting to form an electric net to catch tsetse in flight. Electrocuted flies were recovered from trays coated with adhesive, or by other non-return devices. Both males and females are nighly susceptible to electrocution, and the grids produced an initial knock-down of 96-99%.

The first comparative test with the new method was to compare its efficacy with that of the previously developed sticky surfaces. On the back of a walking man the electric surfaces captured twice as many tsetse as sticky surfaces on the same bait. When this comparison was critically examined the

Fig. 5.3 Catching devices for tsetse flies.
(1) Man with electric net.   (2) Electric decoy (x1.4).
(3) Man with electric surface. (After Vale, 1974c).

possibility of electrified surfaces being attractant in themselves was ruled out, but it was found that the sticky surface, neither attractant nor repellent in itself, became less effective on shiny surfaces of the type which it normally produces. Other possible sources of bias were examined, including the visibility or otherwise of the electric net, and the influence — if any — of the frame and recovery trays. Only when the investigators were satisfied on these points did they proceed to the next phase.

Having established a standard procedure with regard to the model attractant, and the use and operation of electrocuting surfaces, live animal bait was now introduced into the experimental set-up in the form of both domestic stock (mainly oxen) and game animals. For stationary tests, baits were placed in a pen 2.4 m high and 3.6 m in diameter surrounded by a ring of large electrified nets. For mobile bait tests the animals were put in a steel mesh cage or 'crush' mounted on wheels (Figure 5.4). The important question of the attraction of tsetse to host odours in the presence or absence of visual stimuli was tested by two distinct methods. In the first, a model was made by covering the box of a mobile crush occupied by a sheep with

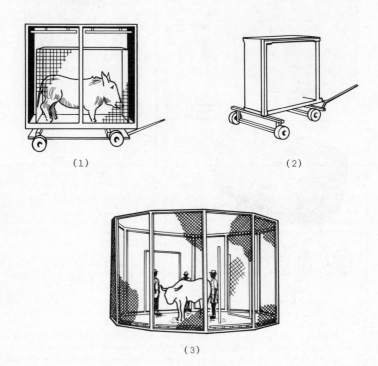

Fig. 5.4   (1) Mobile crush and electric cage used with
a warthog.   (2) Odour model with electric surface.
(3) Electric pen baited with 1 ox and 3 men.
(After Vale, 1974c).

black cloth, through fine holes in which the sheep odour could pass to the
outside. The second design took the form of a ventilated pit (Figure 5.5)
with the bait animal placed in a roofed pit provided with a ventilation
system which blew air from the pit through a small electric net at ground
level.

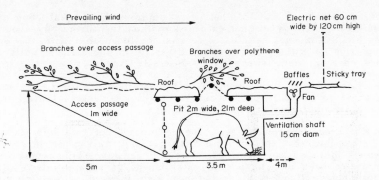

Fig. 5.5 Large ventilated pit used for experiments on attraction of ox odour. (After Vale, 1974c).

In the experiments described, the investigators involved have been careful to point out that while many tests compare mobile and stationary baits or attractants, these really refer to two extremes of what happens in nature where each individual bait animal will be stationary for part of the time and mobile at others. The experiments with electric netting encircling non-human baits under mobile and stationary conditions indicated that stationary baits afford the most representative cross-section of sexes and species, while mobile baits give the best representation of nutritional states. The flies attracted to mobile baits are generally more replete than those visiting stationary baits.

With regard to the relative role of visual stimuli and of odour in determining attraction to bait, not surprisingly the results indicated that visual stimuli are much more important for initial attraction to mobile baits. What was rather unexpected was the finding that odour alone completely failed to attract tsetse to mobile baits. This is in contrast to the attraction to stationary bait where odour alone plays an important part in attracting tsetse to the vicinity of the bait, this orientation being rendered more precise if additional visual stimuli are offered, especially within the close vicinity of the bait. From time to time what appear to be conflicting responses on the part of tsetse to visual stimuli, odour or other stimuli are recorded, and this now seems very likely to be due to the fact that in the course of its life or hunger cycle it passes through a series of phases in which different stimuli are accorded priority, the stimuli which dominate at one phase producing little or no response at another phase.

It was also hoped that all these experiments would throw light on the possibility that the observed differences between men and models could either be due to differences in the numbers of flies attracted from a distance to the vicinity of baits or, alternatively, that the differences could result from closer range behaviour which influences the availability of flies to catching devices. The first thing to determine was whether the depressing effect of man, as convincingly demonstrated in the series of tests with sticky surfaces, was still valid when new catching methods were used. Accordingly, that experiment was repeated using an electric decoy, electric nets and electric surfaces. With all catching devices the model alone caught 2-3 times as many males and 7-20 times as many females as the man alone. The new question this raised was whether the low catch on man was due to low attraction from a distance, or to the fact that flies attracted to the human are somehow discouraged from alighting, or are less available for capture. In order to test this, a comparison was made between the catch on an electric net behind each bait with that of an electric surface on the bait. With the standard model the two catches were approximately equal, but with the human bait the catch on the net exceeded that on the surface by about 3-fold for males and 60-fold for females.   It was concluded that with the model, a high proportion of males and females attracted to the bait actually alight, but in the case of man fewer flies, especially females, show this response after approaching man.   It seems likely, therefore, that despite the very low captures of females in the conventional fly boy round, a swarm of females is still attracted to the bait but because of their low degree of alighting they are not readily available to the catchers.

Direct observation on tsetse.   The combination of relatively large size, distinctive appearance, and daytime activity, have all made it possible to observe the flight and movements of tsetse.   Under favourable conditions many facets of tsetse behaviour have been established, or substantiated, either by direct close observation or at a distance with the aid of a telescope.   By following a number of tame animals, it is possible to record tsetse alighting or engorging on different body regions, and even to follow the movements of individual flies which alight several times on the same animal.   However, the advantages of direct observation at close quarters may be offset by the disturbing effect of the human presence, particularly on female *G. morsitans* and on both sexes of *G. pallidipes*.   In order to be sure that this effect is minimal, observations have to be made at a distance with the aid of a telescope and consequent unavoidable loss of detail.

The complicating visual and olfactory effect of the human presence on tsetse has, however, been successfully eliminated in the case of stationary ox bait by means of an ingenious ox-baited observation pit (Hargrove, 1976). Figure 5.6 shows how the human observer can watch the bait ox very closely through a one-way glass screen.   At the same time a ventilation system ensures that all body odours are effectively removed to an air outlet 50 m away.

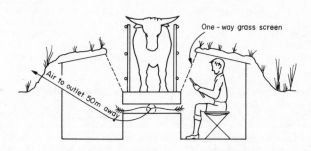

One - way grass screen

Air to outlet 50m away

Fig. 5.6   The pit used for observing tsetse behaviour in the absence of the smell and sight of man.   (After Hargrove, 1976).

Adults of *G. morsitans* and *G. pallidipes* were observed as they fed or settled on the ox, either with the ox alone as attractant, or with ox accompanied by two men standing next to its hind quarters.   These two species could be relatively easily distinguished by the observer on the basis of size, but distinguishing between sexes was not possible.   While the actual visual count of all flies alighting is liable to human error, and less efficient than trapping by means of electric screens, nevertheless the observer was able to keep a separate record of those flies which settle head-down and those that perch head-up.   These two positions indicate different nutritional states of the fly.   Those settling head-down generally do not feed, but may remain on the ox for extended periods. The head–up flies move about less, and a high proportion feed or attempt to feed. Results showed that in the absence of the men approximately four times as many *G. morsitans* and six times as many *G. pallidipes* were recorded perching and feeding on the ox as when the men were present. In the case of *G. pallidipes* at least, in which the high proportion in the head–up position indicated that nearly all flies alighting had come to feed, the presence of man is actually capable of repelling very large numbers of flies which would otherwise have fed on the ox bait.

v.   Feeding Responses to Stationary Hosts

One of the longstanding problems in tsetse behaviour is the question of what proportion of tsetse actually engorge after visiting hosts. This has a bearing on many epidemiological problems, on the diet of tsetse, and on their abundance in relation to the availability of food.

The development of the electric fence or electric net has enabled this question to be tackled afresh from a controlled experimental angle (Vale, 1977). A bait animal is placed in a hexagonal pen surrounded by a complete or incomplete ring of electric bets, usually for a 3–hour period before sunset. The nets measured 3.3 m x 1.5 m, and could be arranged either lengthways or upright. The arrangement with incomplete netting is shown in Figure 5.7. This was designed to trap and stun a sample of the tsetse approaching the bait from outside, and also to deal in a similar manner with samples of tsetse leaving the bait. Flying tsetse do not see these fine nets in time, and when they collide they are killed or stunned and fall into a sticky tray at ground level. In contrast, visual and olfactory stimuli readily pass through the net.

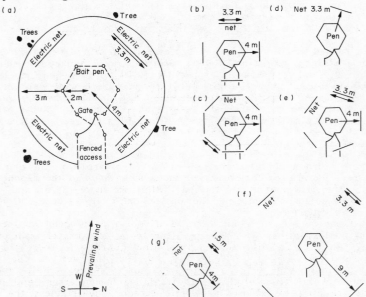

Fig. 5.7   Net arrangements A–G.   Arrangement A is shown at twice the scale of other arrangements.   (After Vale, 1977).

Various possible interfering factors were carefully checked, including the possible visual or obstructive effect of the frame struts or suspensory equipment. In addition, particular attention was paid to the important question of tsetse flight paths in relation to the height of the intercepting nets. This latter point was tested by placing against the upright net (effective height 3.2 m) a series of narrow trays made of transparent fibreglass to intercept and recover flies electrocuted at heights of 20-94 cm, 95-160 cm, 170-244 cm and 245-320 cm. These were in addition to the wide tray at ground level which recovered all flies below 20 cm. All trays were treated with sticky adhesive. The results showed that a very high proportion of the fed flies were caught below 95 cm, confirming previous visual observations by many workers that, after feeding, the flight path of tsetse is low and towards the ground or low vegetation. The results also showed that the flight of unfed flies leaving the bait maintained much the same height as towards the bait. From the rate at which unfed flies decrease with increasing height, it appears that only a few unfed flies travel at a higher level than 3.2 m, and this is supported by observations made at greater heights elsewhere.

These observations on height of flight path were a necessary preliminary to ensure that really valid samples of approaching and departing tsetse were successfully intercepted, and that any loss, other than through the known and measured gaps between the electric nets, was negligible.

Single mammals of ten species were exposed in this experimental layout over a series of exposures. The two important figures required were, firstly, the number of flies that initially approach the bait and, secondly, the proportion of those flies that feed.

Considerable variations in attraction, in which size would appear to be an important but not decisive factor, were recorded. Donkey and kudu, for example, provisionally grouped among the larger mammals, recorded a mean of 494 flies as compared to a mean of 487 for the five smaller mammals: sheep, goat, impala, bushpig and bushbuck. The high catches of 750 and 926 recorded on the two largest mammals, ox and buffalo respectively, were exceeded by the catch of 1081 on warthog, an animals only about a quarter the size.

With regard to the proportion successfully feeding, the differences are much more striking, with the highest proportion on donkey, kudu and ox (0.35 to 0.51), and the lowest proportion (0.00 to 0.02) on impala, goat, sheep and bushbuck. The finding that there may be a wide gap between the attraction of a particular animal to tsetse, and the extent to which tsetse can successfully feed on that host, is strongly influenced by host behaviour. All the smaller animals with low tsetse feeding rates are very restless and intolerant to insect attack, contrasting with the more placid or complacent behaviour of ox or donkey (Vale, 1974a).

This series of experiments also provided a further means of clarifying the already recorded depressing effect of man's presence on the attraction of ox to tsetse. Where ox alone was compared with three accompanying men (Table 5.3) the presence of men halved the attraction of the ox and, furthermore, had the effect of reducing the proportion of flies successfully feeding by about three-quarters.

Table 5.3 Numbers of *G. morsitans* and *G. pallidipes* attracted to, and successfully feeding on, ox alone and ox accompanied by three men. (6 replicates) (After Vale, 1977)

	Numbers attracted	Proportion feeding
Ox	133	0.52
Ox + 3 men	72	0.12

When the experiments with ox and electric fences were extended to cover all months of the year, they embraced seasons characterized by wide differences in tsetse abundance. In general the proportion of tsetse successfully feeding was at its lowest at the season of greatest fly density, the warm dry months of July to September, and rose to a peak in the late rains and early dry season of March to June. For both sexes of *G. pallidipes* as well as for female *G. morsitans*, but not males, there appears to be an inverse relationship between the density of tsetse at bait, and the proportion successfully feeding. This implies that high fly densities may reduce the readiness with which tsetse engorge.

In order to check this possibility under more controllable conditions an experiment was designed in which the numbers of tsetse attracted to bait could be manipulated. With the bait pen and nets arranged as before (Figure 5.7) the ox was stationed in a pen located above the odour outlet of a large ventilated pit (Figure 5.5) in which six oxen were hidden below ground, after the manner of experiments previously described (page 92). The ventilation could be controlled so that the ox could be used with or without the supplementary odour of the six additional oxen. The results (Table 5.4) showed that the extra attraction of the odour from six oxen produced increases in tsetse of the following order: a 2.4-fold increase for male *G. morsitans*, 2.9 for female *G. morsitans*, 4.5 for male *G. pallidipes* and a 4.8-fold increase for female *G. pallidipes*. For both species, the increase in the numbers attracted by the additional odour of six oxen is accompanied by a reduction in the proportion of both sexes which manage to feed.

Table 5.4 Comparison of tsetse (*G. morsitans* and *G. pallidipes*) attracted to ox alone and to ox plus odour of 6 concealed oxen, with regard to numbers attracted and proportion feeding. (After Vale, 1977).

		Ox alone		Ox + odour of 6 oxen	
		Number attracted	Proportion feeding	Number attracted	Proportion feeding
G. morsitans	Males	13	0.32	31	0.23
	Females	33	0.39	95	0.30
G. pallidipes	Males	121	0.51	596	0.39
	Females	235	0.53	1134	0.31

In applying these experimental findings to what might be expected in a natural environment with mixed game and domestic stock, it was estimated that in agricultural areas where oxen and dinkeys predominate, up to 50% of tsetse attracted to these animals might feed successfully. In the case of mixed game animals, however, perhaps only 10-20% of visits to hosts would result in engorgement, and consequently many visits to such hosts might be necessary before each fly could feed successfully.

At this stage the trend of research was increasingly oriented to the practical possibility of controlling, or at least surveying, tsetse populations by means of odour attractants that concentrated flies near traps or sterilizing devices. The odour of a single ox has been found to attract up to 1500 *G. morsitans* and *G. pallidipes* in the course of an afternoon, and odours from six oxen have attracted up to 4000 flies. In order to investigate further the effect of odour concentration, a roofed pit was constructed in which livestock were placed. Odours from the pit were drawn through ventilation shafts to be evacuated at ground level 40 cm upwind of a vertical electric net through which tne odours were deflected. As before, tsetse unable to see the net in time collided with it and were stunned or electro- cuted. In order to concentrate the tsetse attracted from a distance by the escaping odours, a visual bait in the form of a black cylinder was placed just upwind of the net (Hargrove and Vale, 1978).

Cattle (oxen, bulls, cows and calves) were used to produce bait masses of 500, 3500, 6500 and 9500 kg. A maximum bait mass of 11,500 kg was obtained by a combination of 37 cattle, 22 goats, 43 sheep, a donkey and a buffalo. The results are charted in Figure 5.8 and show very convincingly that catches of all tsetse, except male *G. morsitans*, increased considerably with bait masses in excess of 3500 kg.

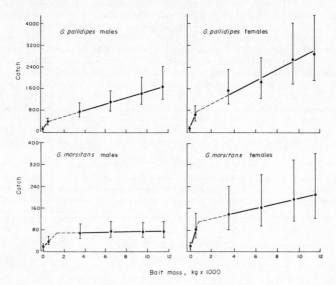

Fig. 5.8   Relationship between bait mass and catches of tsetse.   Vertical lines through each mean indicate the 95% confidence limits.   (After Hargrove and Vale, 1978).

Exactly the same technique of using bait in ventilated pits, combined witn electric nets at the odour outlet, was used for further analysis of the previously established depressing effect of humans on tsetse attracted to ox (Vale, 1979). Men had already been revealed as playing a very incongruous role as baits or attractants to tsetse; on the one hand, moving men attract many males and females of *G. morsitans* and *G. pallidipes*, but on the other hand men can depress the numbers of flies attracted to baits they accompany, and inhibit alighting and feeding responses.

By an ingenious system ot vents, shafts and valves, odours from men or from oxen in separate pits could be delivered to an electric net at ground level (Figure 5.9).   In this way it was possible to compare whole ox odour alone with various combinations such as ox odour plus whole human odour, ox odour plus human breath, and ox odour plus human body odour.   These experiments showed that while a slight reduction in the tsetse catch was produced by adding human breath to the ox odour, the maximum repellent effect could be achieved by the addition of human body odour. The intriguing question of whether or not the repellence is due to a single dominant

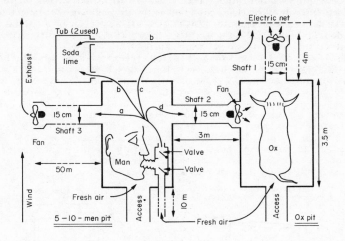

Fig. 5.9 мen pit and ox pit (not to scale). Not all
of the accessories shown were used at any one time.
(After Vale, 1979).

chemical subtance, or to a combination of chemicals, remains one that will
probably take much longer to solve.

iv.   **Natural Resting Sites and Artificial Refuges**

Tsetse flies caught or trapped by the various methods discussed so far
represent the active part of the population at any one time.   There is also
an inactive part of the tsetse population, not sampled by such methods,
which makes use of various natural shelters and resting sites.   These are
mainly flies at stages at which they are not immediately interested in seeking
bait, and also those which are avoiding extreme climatic conditions during
the heat of the day.   A great deal of information is now available regarding
these resting populations for the main species of tsetse.   Some of the information has been gained by direct search for tsetse on tree trunks, branches
and vegetation, but by far the most productive method has involved marking
captured flies in such a way that they can readily be detected in their
night-time resting sites.   Marking methods include the use of fluorescent dyes
detectable by UV light, the use of reflecting patches or the use of marker
based on microscopic glass beads, both the latter being detectable by torch
light.   All these methods have the advantage that marked flies can be
detected in situ without actually capturing them or disturbing them.

These methods have helped to define the nature of the resting sites,
the vertical distribution on tree trunks and on branches, and also differences
between daytime and night-time preferred sites.   They have also disclosed
the existence of seasonal variations in resting behaviour, and also changes
in the preferred sites during the course of the day.   The composition of the
resting population with regard to species, percentage of males and females,
and physiological condition has been compared with that obtained by other
sampling methods, an aspect which will be discussed further later in this
section (page 106).

Accurate information about the use of different resting sites and surfaces, and about the movements and behaviour of tsetse in relation to these
preferred sites, is important in practical tsetse control by insecticide application to tne particular type of bush, woodland, or forest frequented by
tsetse.
The question of tsetse behaviour in relation to resting sites is another
facet of tsetse ecology which has been the subject of critical examination and
experimentation in Rhodesia.   The basis of this work has been the use of
artificial refuges of various types which have been designed with regard to
the characteristics of the preferred resting sites in nature (Vale, 1971).   The
range of refuges found attractive to tsetse are shown in Figure 5.10, the
important features common to all being the shaded entrance to the refuge,
and the provision inside the refuge of dark cool recesses or cubby holes.   In
the nut refuge, tsetse were found to rest mainly in the cubby hole, and to a
lesser extent on the rear wall of mud.   In other artificial refuges tsetse
rested almost exclusively in the drums or pipes.

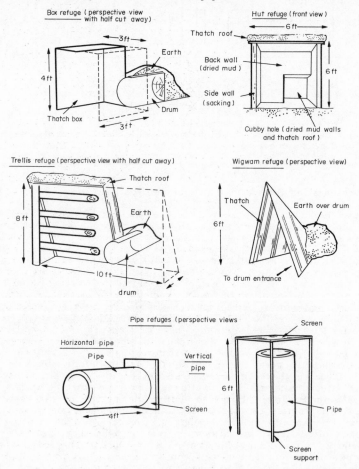

Fig. 5.10   Some artificial refuges.   (After Vale, 1971).

Tsetse resting in such refuges could often be counted in situ, and samples caught by hand net for identification or examination. However, it was seen that some tsetse were disturbed by capture activities or by the collector himself, and it was soon realised that modifications in capture technique would be necessary. In the case of box and hut refuges, curtains were fitted which could be triggered to close from 30 yards away. The approaching collector then first made sure that any tsetse attracted by his presence were net-caught before he crawled under the curtain and captured all tsetse enclosed in the refuge. The curtain was then re-opened and the trigger re-set.

Another modification was to reduce the entrance to the trap to a narrow slit, and arrange the design so that tsetse leaving the shelter after they had rested in the drum were guided into a small flask (Figure 5.11) provided with Vapona pest strip which rapidly killed them.

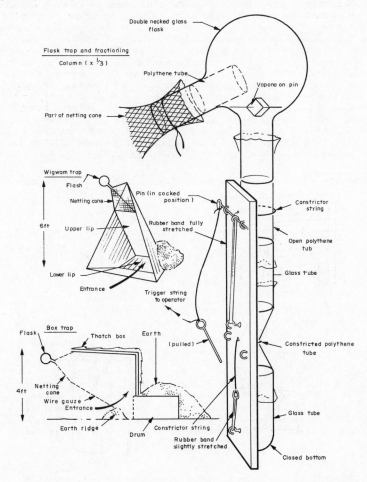

Fig. 5.11 Some trapping systems. The fractionating column is shown with two segments but was used with seven. (After Vale, 1971).

In order to study the timing of exits, a system was designed to frac-
tionate entries into the flask.   Flies entering the flask were treated by
Vapona and fell to the base of the flask and down a tube of alternating
glass and polythene tubes.   Segments of the column could be isolated by
pinching the polythene tubes successsively from the base of the column to the
top and so separating samples emerging at different times.   The closure of
the polythene tubes was triggered by a string from 30 yards away.

To trap tsetse flying into a box refuge, the front of the refuge was
covered by a screen of fine netting, and an electric grid fixed just outside
the net.   Tsetse attempting to enter the refuge were killed or stunned and
fell into a sticky trap on the ground.

The effect of location was studied by siting box refuges in different
types of vegetation from fringing riverine bush to deciduous woodland.   The
results showed that in all months of the year except August the riverine
vegetation recorded much higher catches.   In order to check whether the low
catch of tsetse in the box refuge in deciduous woodland indicated a low
preference for those on the part of the tsetse, or were really a reflection of
low tsetse populations in that habitat, additional methods were used to
estimate the existing tsetse population.   By using Morris traps and ox bait
captures it appeared that high tsetse population densities could occur in
deciduous woodland, indicating that many of the flies active in that habitat
moved to the nearby riverine vegetation to seek shelter, even though suitable
refuges were available on the spot.

By making two-hourly inspections of artificial refuges in the form of
box shelters, and natural resting sites in the form of rot holes, a pattern of
occupation of these places by tsetse in the course of the day was defined
(Figure 5.12). This showed a peak of occupation around 14.00 hours, a time
of day when temperature, saturation deficit and light intensity were very
high.

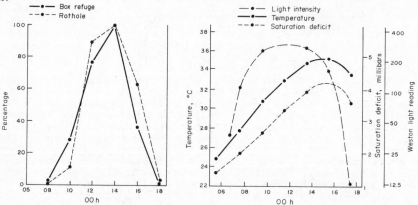

Fig. 5.12   (Left)   The   total   tsetse   seen   at   each
inspection of a box refuge and a rot hole as a percentage of those seen at 14.00 h. (The totals seen at
14.00 h in the box refuge were 341 and in the rot hole
99);   (right)   the   diurnal   course   of   meteorological
conditions.   (After Vale, 1971).

Both sexes and species of tsetse showed similar patterns of refuge
behaviour.   Entrance started in the mid-morning as temperatures approached
32°C, reaching a peak at 10.00 to 14.00 hours.   A few exits took place
before 14.00 hours, several from 14.00 to 16.00 hours, and a marked exit

peak occurred from 16.00 to 18.00 hours. The period of highest refuge entry coincided with the period of lowest capture of active tsetse by ox bait and Morris trap, both of which demonstrated early morning and late afternoon peaks.

vii. Comparison of Sampling Methods

Much of the knowledge of tsetse behaviour and ecology has been built up over the years from sampling data obtained by a wide rangte of capture techniques. At the same time it has long been realised by all concerned in this work that samples differ in numbers and composition according to the mode of capture used. The problem of correctly interpreting capture data in terms of real tsetse populations has provided a challenge and a constant stimulus to devise new techniques and to re-examine old ones. All the current capture and sampling methods have certain deficiencies and biases, either with regard to representation of species, proportion of the sexes, nutritional state or age representation. In order to quantify these specific biases, a great deal of work has been done in which different sampling methods have been carried out concurrently in the same area, under similar conditions of climate and vegetation. Some earlier and well-recorded examples are as follows (Muirhead-Thomson, 1968): a continuous comparison throughout the year of Morris trap ('animal' trap) versus conventional fly round in both West Africa and East Africa; a comparison of cattle bait, Morris trap and fly round in Uganda; a comparison of four sampling methods in East Africa, viz. fly round, ox bait capture, resting tsetse in natural sites, and vehicle catch (flies attracted to slow-moving vehicles). In another investigation the addition of three more techniques to the latter (viz. animal trap, moving ox bait and stationary men) produced a valid and valuable comparison of seven different sampling methods for *G. morsitans* and *G. fuscipes*. This last comparative study was done at a time which roughly coincided with the intensification of tsetse research in Rhodesia in the late 'sixties, and could be taken as an important milestone or turning point in the approach to sampling problems.

Since that time the development of new study methods involving the use of sticky trapping surfaces, electrocuting grids, ox odour attractants and artificial refuges has led to a new spectrum of techniques available for comparison experiments (Hargrove, 1977). At the same time the constant search for new mechanical traps for tsetse has continued, and a great deal of ingenuity has gone into various modifications and improvements in design, all based on the common principle of attracting tsetse to a model or bait in such a way that the flies are funnelled through a series of net cones of baffles into non-return cages. To the long list of traps named after their respective designers, such as the Harris trap, the Chorley trap, the Morris trap, etc., the most recent trends and improvements are best exemplified by the vertical vane traps studies in East Africa and Rhodesia (Hargrove, 1977), and the biconical trap designed by the French workers in West Africa (Challier and Laveissiere, 1973; Laveissiere and co-workers, 1979) and subsequently adopted in Nigeria (Koch and Spielberger, 1979). Both traps, designed quite independently in different parts of Africa, have several essential features in common. The main components of the vertical vane trap are (a) a model, or visual bait to attract tsetse to the trap, (b) a condenser, or non-return device to restrict flight away from the bait, (c) a collector or removable box to retain the flies in a compact volume, and (d) a hat to concentrate the flies below the collector (Figures 5.13, 5.14).

The biconical trap designed by the French workers in West Africa is based on the same principle that tsetse in flight are attracted to contrasting dark and light surfaces, and that once they have entered the dark interior

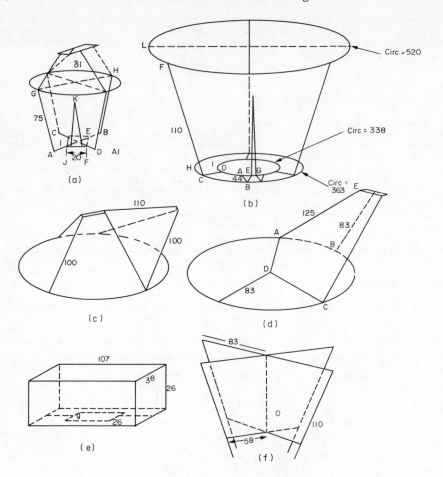

Fig. 5.13 The components of the vertical vane traps:
(a) Series A trap without collector;   (b) V44 condenser;   (c) Hat H2;   (d) Hat H4 (only one side arm
shown);   (e) Collector C5;   (f) M3 model (M2 had the
same shape and base length but the edge was reduced
from 100 cm to 55 cm).   (After Hargrove, 1977).

their subsequent movements are upwards to light, where a non-return exit
confines them to the upper part of the trap.   The French model emphasises
the dark lateral shady openings designed to intercept the direct flight of the
fly, so that attracted tsetse do not have to enter the trap from underneath
(Figures 5.15, 5.16).   This trap was designed for use in riverine areas
where *G. palpalis* and *G. tachinoides* are dominant.   Comparisons between
biconical traps and standard fly boy captures showed that whole day captures in the trap were generally as effective as the fly boys for *G. tachinoides*, but rather less effective in the case of *G. palpalis*.   The particular
merit in the case of *G. tachinoides* is that the trap captures a much higher

Fig. 5.14   Condenser V44 with hat H4 and collectors C4.
(After Hargrove, 1977).

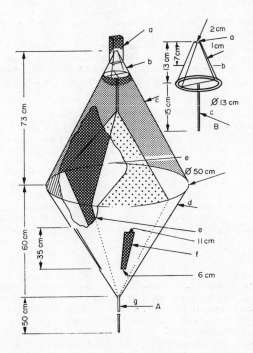

Fig. 5.15   Diagram showing different components of the biconical trap.
(After Challier and Laveissiere, 1973).

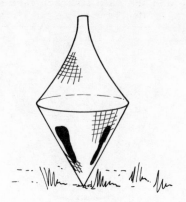

Fig. 5.16   Biconical trap in position on the banks of
the Leraba River, Ivory Coast.
(After Challier and Laveissiere, 1973).

proportion of females than the fly round.   The lower inverted cone of the
trap, originally of white material, has now been replaced by blue material,
found to be more attractive.   This new trap has now been adopted on an
extensive scale as a method of tsetse control and population reduction, tsetse
which are attracted to and enter the trap being automatically killed by
insecticide (decamethrin) previously applied to the trap.

To return to the new approach to the comparison of current sampling
methods, again this is best illustrated by experiments carried out in Rhodesia
on the following lines (Phelps and Vale, 1978;   Vale and Phelps, 1978).   In
a small block of riverine woodland the following techniques were used concurrently every day for a week: (a) three ventilated pits baited with an
ox;   (b) three mobile baits operated separately on a single path, the baits
being ox with three men who used hand-nets to catch flies alighting on the
ox;   (c) three men catching flies alighting on a black cloth screen carried
by two other men;   (d) a similar sized screen carried by two men and provided with an electrifying grid;   (e) artificial refuges comprising 16 box
refuges, 12 horizontal pipe refuges and a ventilated pit without oxen;   (f)
natural refuges comprising boles of trees and rot holes, and (g) tsetse
resting on horizontal branches.

In addition to recording sexes and species in each catch, particular
attention was paid to the nutritional status of flies sampled in different
ways.   All males recovered in good condition were hunger-stages as teneral
(young soft flies which have never fed), Stage I (recently fed flies with
abundance of undigested blood) through stage II and III to stage IV (old
flies with no residual blood meal and little fat).   A more precise index of
nutritional state was also obtained by measuring the weight of fat and of
undigested blood in each fly.

Allowing for day to day fluctuations in catch, and for the fact that
temperature conditions throughout the experiment were not uniform, the
salient features of these tests is shown in Table 5.5, the comparison of
samples from different sources being referred to the existing composition of
the whole natural population as calculated by methods described later in this

Table 5.5   Salient features of the composition of the wild population of
tsetse and the composition of various catches
(After Vale & Phelps, 1978)

	% females *Glossina morsitans*	*Glossina pallidipes*	% *G. pallidipes*	% teneral
Wild population	65	75	90	5
Artificial refuges	60	50	85	5
Ox-baited pits	60	70	90	5
Mobile ox	20	40	75	15
Mobile screen, hand-net	5	45	15	30
Mobile screen, electric	5	60	60	30
Branches	35	35	90	30

section (page 108).   In terms of the proportion of each sex and species, and
of the proportion of tenerals, the pit and refuge catches appear roughly
representative of the wild population.   Refuge catches have the additional
advantage of capturing flies in a wide range of nutritional states likely to
reflect a truer picture of the nutritional status of the population at large.
However, it is necessary to qualify this general statement in the case of
G. pallidipes in which the composition of the refuge catch is much more
dependent on temperature than in the case of *G. morsitans*, and it is only at
the highest temprature experienced (41°C) that the refuge collection of
G. pallidipes equates to that of the wild population as a whole.
It will also be seen that the percentage of tenerals in refuges is very
low, as in the wild populations, as compared for the high proportions taken
in mobile screens. Furthermore, it was only in the refuge and resting catches
that freshly engorged flies were taken in high proportions, in accordance
with that of the wild population.   All three types of catch from mobile bait
appear heavily biased in terms of absence of certain nutritional groups, in
the high proportion of tenerals, and in certain aspects of sex and species
representation.
Examination of the figures in Tables 5.5 and 5.6 also revealed that the
concept of 'resting' populations needs revision, and that this is by no means
uniform in composition.   For example, the proportion of females in pit
refuges was much greater than in natural refuges or in other types of
artificial refuge.   The catches from natural refuges were much higher than
those from branches.   Moreover, the composition of the two types of catch was
distinctly different, the refuges producing relatively high proportions of
G. morsitans, and high proportions of females of both species (Table 5.6).
For both species the highest proportion of females (70-74%) were
recorded from pit refuges as compared with other refuges.   The conclusion
from all this is that all refuge collections cannot be taken as equally
representative of the wild population at large.

Table 5.6   10 day catch of tsetse from branches and natural refuges.
(After Vale & Phelps, 1978).

	G. morsitans Males	Females	*G. pallidipes* Males	Females
Branches	26	15	284	166
Natural refuges	152	240	544	572

viii. Population Studies on Tsetse

The new spectrum of sampling *G. morsitans* and *G. pallidipes* by a variety of methods, and the critical assessment of the composition of captures of both active and resting fractions of the population, provided a stimulus to re-examine old problems of tsetse population density.

The calculation of population density, and establishing the composition of tsetse populations in their natural state, by means of mark-release-recapture data played a very prominent part in earlier tsetse studies in East Africa. Because of the capture methods used at that time, in which males predominated, calculations were based essentially on marking, release and recapture of males. When this problem was re-opened in Rhodesia, there were several innovations in the design of experiment (Phelps and Vale, 1978). Firstly, care was taken to ensure as far as possible that the experimental area was an isolated one in which immigration and emigration of flies would be minimal. In this case the isolation of the 13.5 ha area in the Zambesi valley was achieved by selecting a time of year in the hot dry season when the woodlands surrounding the area were leafless and unattractive to tsetse. In contrast the riverine experimental area had trees providing dense shade.

The second important feature was that all the tsetse which were going to be marked and released were collected in refuges, both natural and artificial, thus ensuring a good representation of sexes and nutritional states. Tsetse marked with a paint spot were released inside these refuges in such a way that they left the refuge naturally when the curtains were opened. All flies were marked and released on one day.

The third important feature was that in the daily captures which were made 10 days following release, the variety of capture methods described in page 106 were used, again ensuring a wide spectrum of sampling. Estimates of populations of both sexes and species were based on the combined catches of non-teneral flies taken by various sampling methods. The numbers of both species marked and released, and the proportion of those which were recaptured, are shown in Table 5.7.

Table 5.7   Mark-release-recapture experiment. (Phelps & Vale, 1978).

| | *G. morsitans* | | *G. pallidipes* | | Total |
	Male	Female	Male	Female	
Flies marked	107	60	565	665	1230
% recaptured over 10 days	24.3	16.7	21.8	12.0	17.1

Total flies captured over 10 days by all methods = 43,614 non-tenerals
2,651 tenerals

The estimations of population density were calculated by several different standard procedures along lines well documented in standard ecology textbooks (Southwood, 1980). The fact that six different mathematical approaches produced reasonably similar estimates of population numbers suggested that the final figures were a fair representation of the true situation. Within the study area the calculated figures of non-teneral fly density came roughly within the following ranges:

G. morsitans	males		7,000 - 10,000
	females		15,000 - 19,000
G. pallidipes	males		52,000 - 57,000
	females		140,000 -168,000

and from this was derived an estimated total tsetse density of 1.24 flies per m².

Further information was provided on the basis of the amount of blood known to be taken in an average tsetse blood meal, and the period elapsing before food reserves from the blood meal became exhausted. The average period between blood meals was then worked out as 3.7 days for *G. morsitans* and 5.5 days for *G. pallidipes*. These figures were then used, along with the population estimates, to give estimates of the quantity of blood needed to support the calculated population of tsetse, and the number of buffalo or kudu required to supply this quantity of blood on a daily basis. This latter calculation was assisted by available knowledge about how much blood an animal such as an ox can lose in the course of a day without loss of condition.

Within the study area itself it was calculated that a minimum of eight buffalo or 25 kudu would be required to support the estimated tsetse population. Analysis of blood meals of tsetse captured from refuges at the end of the experimental period showed that a wide range of hosts were available and were being fully utilised by the flies present. This fact was considered to provide further grounds for regarding the estimates of tsetse density as being realistic.

Perhaps enough has been said about this aspect of tsetse research to illustrate the way in which a solid groundwork of information about the advantages and limitations of each capture method employed offers the best chance of venturing safety into this very difficult, controversial and often speculative field of animal population studies.

Tsetse population studies in Uganda. An interesting contrast to the previous studies on populations of the savannah species *G. morsitans* and *G. pallidipes* has been provided by similar investigations on a typical thicket and forest species, *Glossina fuscipes* (Rogers, 1977). In approaching fundamental problems of movement and distribution of tsetse, the investigators in Uganda have preferred to use classical sampling and capture methods. This work was carried out on the north-eastern shore of Lake Victoria, in Uganda, during the dry season of January to April 1971, and was designed to test the feasibility of estimating population numbers in a relatively large area with virtually unrestricted fly movement.

A point was chosen in the middle of the experimental area on the basis of its apparently high abundance of *G. fuscipes*. Flies approaching a stationary party of three or four humans were captured on members of the catching team or on the ground in the immediate vicinity. The flies, captured with hand nets, were marked individually with oil paints and released immediately. For recapture purposes a series of decagonal fly rounds were cut through the bush at radial distances of 300, 400, 600, 800 and 1000 yards from the point of release, the arrangement forming five 'concentric' fly round round the point of release (Figure 5.17). The five fly rounds were each marked out into catching stations at 100 yd intervals over which a party of two collectors carried out conventional captures once or twice a week. During this study, marking of flies at the centre of the test area was carried out for 2-3 hours each weekday morning, while fly rounds were worked in the afternoon.

During the study period 2298 male and 1942 female *G. fuscipes* were marked and released at the centre of the fly rounds. Out of a total of 8022 males and 3478 females taken in the fly rounds, there was recapture of 55 males and 28 females at the centre, and 77 males and 10 females in the concentric fly rounds. In the standard fly rounds, the tsetse population attracted to human bait when moving contained a high proportion of males. When the bait remained static, a less biased sample of the population was obtained, with the female catch proportionately higher than in the mobile bait.

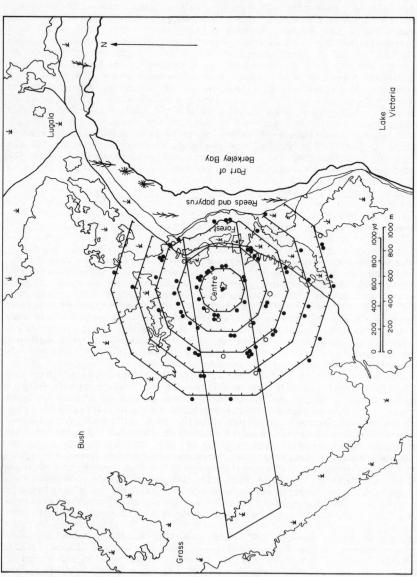

Fig. 5.17 Plan of the study area in the South Busoga forest, showing the point of marking and release at the centre of the five decagonal fly rounds. Parallel lines indicate the track of a pre-existing fly round. Recaptures at each sector on the fly rounds are indicated: ● male, ○ female. Lugala is a small fishing village at approximately 33°54′E, 0°12′N. (After Rogers, 1977).

The analysis of male recaptures provided information about the feeding interval, using the intervals which elapsed between release of individually marked flies at a particular stage in the hunger cycle, and their recovery at precisely the same stage. This is essentially the same approach as used in the original classical studies on populations of game tsetse in the 'thirties, and the conclusions reached were much the same regarding a period of 4 days being the normal hunger cycle.

With regard to the information provided by this experiment on the distribution of the adult population, it was found that one of the factors affecting distribution was the amount of light, this in turn being determined by the vegetation cover. There is an intermediate illumination at which the availability of male flies is at a maximum, and this coincided with the thick scrub covering most of the area. Both at lower illuminations (belt of tall forest) and at high illuminations (patches of open grassland) catches were lower.

With regard to the distribution of re-captures, there seems to be no tendency for the flies to move in any particular direction from the release point. Recaptures appear to be distributed quite at random in the fly rounds, and suggest that movement of *G. fuscipes* in this habitat is a simple diffusion from the point of release (Figure 5.17).

ix.   Electric Net Sampling in Other Parts of Africa

The successful development of electric grids and electric traps for tsetse in Rhodesia has stimulated other workers to develop or modify the basic idea for their own requirements. Work in Ethiopia, for example, has been directed towards reducing the weight of equipment — including 12-volt battery — while maintaining the voltage necessary to stun tsetse (Randolph and Rogers, 1978;   Rogers and Smith, 1977).   For use in remote areas it was also necessary to keep the power requirements to a minimum.   This was achieved by means of a capacitor which acts as an energy store, and which is discharged only when the fly lands on the grid and is given a powerful shock.

Trials on a subspecies of *G. morsitans* (viz. *G. morsitans submorsitans*), in which the trap was worn on the back of field assistants moving along a set path and stopping at intervals, confirmed that without the capacitor flies which made only a fleeting visit to the grid were able to escape. This new circuit provided a light and efficient alternative to the Rhodesian design.   Although perhaps slightly less effective in stunning flies, the reduction in combined weight of circuit and 2-V battery to 1 kg had great advantages in portability.

So far all these developments had been carried out with members of the 'morsitans' group of tsetse.   The next noteworthy step was the extension of this work to West Africa and a different group of tsetse, namely the 'palpalis' group, including *G. palpalis* and *G. tachinoides* (Rogers and Randolph, 1978).   The essential difference between these two groups in the present context is that the *palpalis* group has a natural propensity for primate blood.   Flies were caught on stationary and moving men either by catchers using nets, by the electric back pack as designed in Ethiopia, or by an electric screen 60 x 90 cm mounted vertically on four wheels with a towing handle.   The latter two methods were used separately in standing catches, but were used together in some of the moving catches.   In the latter case the screen was pulled immediately behind the wearer of the electric back pack, so that the back pack tended to catch flies preparing to land on the bait above the waist, while the screen sampled flies approaching at a lower level.

The tests revealed differences according to whether the bait was mobile or stationary. During standing catches the electric traps were not markedly superior to the catch by men using hand-nets, but with moving catches the electric traps on average caught higher numbers of both males and females of G. palpalis and G. tachinoides.

One obvious factor affecting the comparison between hand catching and electric grids is the time taken by catchers to remove each fly from the net, sex it and record it. This may not be too important at low fly densities with few flies arriving, but with high densities there may be a limit to the number of flies the catcher can cope with.

Another possible explanation is provided by an experiment in which one party employing one catching method was followed 20 minutes later by another party following along the same track and employing another catching method. On the return journey along the same track, the sequence of the two methods was reversed. It was expected that the first party along the path would deplete the population of flies for the second trapping party, although each party would be expected to catch a fixed proportion of the flies available to it. The results did in fact show that the passage of one catching party along the track did affect the behaviour of some of the remaining flies, making them unavailable to the second party 20 minutes later. However, the effect was not the same for all parties; it was the men with hand nets who produced this effect on the following back pack captures, possibly due to the whisking of the fly nets driving tsetse away long enough to deplete the catch by the party following 20 minutes later.

One other factor which must be taken into account in such comparison trials is also the fact that the back pack traps only those flies alighting on the carrier, while fly boys can catch tsetse slighting anywhere on the body. However, an examination of the figures obtained from back pack and electric screen showed that electric packs which work on part of the body are able to sample most or all of the G. palpalis available to the hand-net. The figures also disclosed a difference in behaviour of the sexes in that male and female G. palpalis had different approach heights to the bait, thus showing similar trends to those observed in Rhodesia with G. morsitans in which the top half of sticky surfaces of both model and man recorded higher percentages of the female catch and lower percentages of the male catch as compared to the bottom half.

The possibilities of new electric trapping devices were also tested out by another group of workers in a different part of Northern Nigeria (Koch and Spielberger, 1979) by comparing them with two other capture methods, namely hand-nets and the biconical trap (page 103). One of the two test areas had a high density of Glossina palpalis, while the other had a low density of G. palpalis and of G. tachinoides.

The electric trap used was a modification of the electric back pack described above, but the size of the catching surface was increased, and the colour changed to dark blue in order to increase its attraction to G. palpalis. In this trial, the trap was carried along the fly rounds on the back of one man with the other collector following on behind and watching for any stunned flies trying to scape. Evenly distributed along the route eight stationary biconical traps were set up at points where footpaths and cattle tracks crossed streams, and at the washing and watering points used by local people. The catching stations where flies were caught by hand-nets were mainly those where biconical traps were sited.

In area I, where fly density was high, 328 G. palpalis were captured, of which 67% were caught in the biconical trap, 21% by hand-net, and 12% on the electric trap. In area II where fly density was much lower and where only 125 G. palpalis were taken, more than half (57%) were taken by hand-net. The electric trap took only 9% of G. palpalis and 15% of G. tachinoides. While these results might appear to be at variance with those reported above by the first team, densities of G. palpalis were much higher in that locality of Northern Nigeria than in area I of the present trial.

Taken in conjunction, the results of the two independent studies indicate that with *G. palpalis* at least, the electric back pack is at its most efficient at high fly densities and relatively inefficient at very low densities.

Some idea of the relative densities in this trial were obtained by hand catching on the fly rounds prior to the trial. In the 'high density' area 10 flies were recorded per kilometer, while in the 'low density' area only 0.3 flies were taken per kilometer. These trials also served to define the advantages and limitations of the three sampling methods under similar conditions. In the case of the modified electric back pack, portability was partly offset by the fact that equipment was liable to mechanical damage in rough field conditions. It was also observed that many flies appeared to be only stunned by the electric grid, and managed to escape from the collecting box.

In the case of the biconical trap, satisfactory performance was slightly marred by its low mobility and by the fact that frequent inspection was necessary to ensure that captured flies were not destroyed by ants and other predators. It was also found that very careful siting was essential to ensure good performance.

Both groups of workers in Nigeria found that electric traps caught a higher proportion of recently-fed flies than hand catching did, and this advantage was also evident over the biconical traps tested by the second team. This is an important feature in view of the fact that recently en-gorged female *G. palpalis* are difficult to detect in their natural habitat at these fly densities, making it difficult to obtain adequate samples for blood meal testing by search alone.

x.   Host Selection

In their biting habits tsetse make their most obvious presence felt when attacking man or his cattle and other domestic stock. But this is only one small facet of the normal behaviour pattern of tsetse males and females seeking a vital blood meal. The wide ranging habitats of tsetse include various types of savannah woodland, riverine forest and thickets, where a variety of animals may abound. This is particularly marked in the case of game tsetse of East Africa, *Glossina morsitans*, *G. pallidipes* and *G. swynner-toni*, many of whose habitats are vast areas with a rich mammalian fauna, some members of which are present in high densities. The dependence of some species of tsetse on these varied sources of blood forms the basis of some of the earlier control measures employing game destruction or game exclusion. But the precise degree of dependence of such tsetse on the blood of wild animals, and the extent to which some tsetse select some species in preference to others, has proved a challenging problem of which much remains obscure and unexplained.

In the earlier days direct observation played an important part in gaining some idea about which game animals appear more attractive to tsetse than others, and in this respect the size and conspicuous appearance of tsetse combined with their daytime activities made such observation possible and rewarding, especially with the assistance of a telescope. Such direct observations disclosed, for example, that some common game animals such as zebra did not appear to attract tsetse in numbers. However, it was not until feeding behaviour was first studies intensively by means of the precipitin test for the identification of blood-meal source, that clearer patterns began to emerge. Such work showed that tsetse feed predominantly on only a few of the many species of mammal available in their environment. Some of the highly preferred hosts such as warthog and bush pig were not conspicuous either for their large size or for their numbers. In contrast, very little feeding appeared to take place on such widespread ungulates as the impala.

The first precipitin test survey, in East Africa, used mainly engorged or partly-fed tsetse attracted to the survey parties carrying out fly rounds in different areas of vegetation. Those flies were predominantly males, with females under-represented, and the samples could well be biased in terms of the tsetse population as a whole. Subsequent precipitin test surveys have endeavoured to base results on a much wider sampling spectrum, with particular reference to gorged or recently-fed flies taken in natural shelters or resting places. Samples from such sources are now recognised as being more representative of the population at large, and less likely to be biased by the presence of a catching party and of alternative hosts.

In more recent years both the precipitin test and direct observation have continued to play a vital role in elucidating tsetse feeding behaviour, but they are being coordinated more critically and supplemented by planned experiments involving known numbers of host animals, both domestic stock and wild game animals. This new approach is best illustrated by the following examples.

The Rhodesian work described earlier in this chapter disclosed two important factors influencing host selection. First was the finding that the presence of humans had a depressing effect on the numbers of tsetse alighting on bait, this being particularly marked for female *G. morsitans* and both sexes of *G. pallidipes*. Second was the disclosure that, despite a low degree of alighting or biting an animal bait or model under these circumstances, the bait could still attract a following swarm of flies. Two separate responses were involved: firstly, the attraction of tsetse to the vicinity of the host; and, secondly, the responses which determined whether this was followed by alighting and biting.

Having established these points, an experiment was carried out in which an observer followed a number of game animals as they wandered — with frequent stops to feed and rest — in deciduous woodland where *G. morsitans* was abundant (Vale, 1974a). Oxen were also watched as they grazed in open areas nearby which were less heavily infested. The results showed that with regard to the numbers of alighting and engorged flies seen on the whole body of each animal, there were many landings and engorgements on the ox, dog, bushpig and older warthog, but few landings and no engorgements on the impala and bushbuck. There was no direct way of checking the actual number of flies attracted to each bait, but previous experiments had shown that animals of the same size attract roughly the same number of tsetse. Consequently, the initial attraction of tsetse to impala, bushbuck, dog, bush-pig and older warthogs, all of which were about the same size — including the observer — would not differ substantially.

All impala and bushbuck were observed to be very intolerant to tsetse attack, responding by leg kicks, tail flicks, head movements and skin 'rippling'. The separate records of numbers alighting, followed by observations to see how many of those led to successful engorgements, enabled calculation to be made of the number of landings for each engorgement. For the favoured hosts, ox and older warthog, the numbers were 26 and 30 respectively; for the dog it was 140, bushpig 380, and for the impala and bushbuck the number was infinitely large.

In this experiment all the figures for biting rates were very low, being only 3 engorgements per hour for the favoured older warthog. This, of course, refers only to the one side of the animal visible to the observer. Nevertheless, from previous findings it seemed likely that the presence of the human observer did depress the numbers alighting and feeding, and that much higher biting rates could be expected under natural conditions. These observations still left several problems unsolved. The low attraction of *G. morsitans* to bushbuck contrasts with experience with other species of tsetse, bushbuck accounting for nearly half the bovid blood meals recorded, or even providing the bulk of the diet, in the case of *G. pallidipes*. No doubt other factors such as density of bushbuck relative to other available

animals, and the extent to which different species of tsetse are attracted to
stationary rather than moving bait, may eventually explain these discrepancies.

The role of different wild animals in supporting tsetse populations has
been assessed by a different experimental approach involving selective game
removal (Vale and Cumming, 1976). In a wildlife research area in Rhodesia
two experimental blocks, 10.9 km^2, each with abundant game population and
variety of vegetation, were designated, one as a control and the other as an
area of warthog removal. Information about tsetse populations was obtained
by conventional fly rounds, and by stationary ox bait. Replete flies caught
by fly round parties and gorged flies found on vegetation provided samples
for precipitin tests of blood meals. By trapping and shooting, most of the
warthog had been removed by the end of two months, prior to which activity
a warthog-proof fence had been erected round the block. The removal of
warthog was followed subsequently by removal of all elephants.

Prior to warthog removal suid blood made up 77% of identified blood
meals in the trial area and 84% in the control, more specific tests showing
that the bulk of suid blood meals were indeed from warthog. During and
subsequent to warthog removal, this figure fell to 10% in the trial area.
Over the same period the percentage of bovid blood in identified blood meals
increased from 16% to 77%. The bulk of those blood meals were from kudu,
whose numbers in the trial area had not been affected. There was also a
marked increase in elephant blood meals until those animals also were
removed.

With regard to the effect of warthog and elephant removal on the tsetse
population, there was no significant reduction in abundance of tsetse as
judged from males taken in fly rounds. On the other hand, the rate of
tsetse blood meal collection from resting places indicated a halving of population. This, together with an increase in the proportion of females taken in
the warthog-free area, suggests that the population was under some stress.
If the trial area had been more completely isolated and remained completely
free of warthog for a longer period more drastic effects on the tsetse population might have occurred. There is evidence that if such isolation is
ensured, as on an island in Lake Kariba, removal of warthog — in this case
a spontaneous decline — can have a devastating effect on tsetse, even though
other animals in the form of impala, sable and baboon are present.

The use of the precipitin test in elucidating tsetse reactions in experiments involving controlled numbers and species of hosts is also well illustrated by a rather different situation in Rhodesia (Pilson and co-workers,
1978; Boyt and co-workers, 1978). In the transmission of animal trypanosomiasis it has been found that in the areas denied to cattle because of the
disease, sheep and goats are able to exist. The question therefore arises as
to whether sheep and goats are less attractive hosts for tsetse. The problem
was studied by two methods: firstly by directly counting flies attracted to
and attempting to feed on bait, and indirectly by identifying the source of
blood meals in recently gorged tsetse (Pilson and co-workers, 1978). In
preliminary experiments with stationary bait animals five choices were offered:
ox, ox plus sheep, ox plus goat, and sheep and goat, the animals being
tethered under a shady tree in riverine vegetation in the Zambesi valley.
The same choice was offered in experiments with mobile baits covering identical traverses. As the baits were accompanied by two men — catcher and
recorder — the effect of human presence was checked in a separate experiment
in which three tethered oxen were observed by three teams of 3, 6 and 12
men working in rotation. In an extended experiment covering four seasons of
the year, the relative attraction of ox, sheep and goat on their own was
compared at peak biting periods appropriate to that particular season.

In the preliminary experiment 566 feeds were recorded on the ox alone,
498 on the ox when providing a bait with the sheep, and 500 on the ox when
associated with goat. No feeds were recorded on either goat or sheep when

Table 5.8   Relative attraction of ox, goat and sheep to tsetse.
(After Pilson et al., 1978).

	Number of G. morsitans		Number of G. pallidipes	
	Males	Females	Males	Females
Ox	211	103	427	237
Sheep	9	2	1	0
Goat	5	0	0	0

combined with ox.   Five flies fed on sheep alone, and one one on goat alone.
The results of the main experiment with individual animals is shown in
Table 5.8.
These experiments demonstrate clearly the overwhelming attraction of
oxen to tsetse as compared with sheep and goats.   Many of the flies attracted
to goat and sheep were prevented from biting by vigorous host defensive
actions.   With regard to the effect of increasing numbers of humans, the only
significant effect was seen in the case of G. pallidipes where the alighting
response was progressively depressed, with an even more marked depressing
effect on feeding.
In a second series of experiments an area was divided into four sectors
in each of which cattle, sheep, goats and donkey were grazed separately in
rotation.   On each day following exposure the area was searched for engorged
tsetse.   A total of 1412 blood meals were collected: 512, 22 and 46 were
positive to ox, sheep and goat, respecitvely, confirming that in the absence
of the human collector the ox has the dominant attraction.
An extension of this grazing design of host selection experiment was
motivated by the finding that under certain conditions game tsetse can be
be supported by equids (in this case donkeys), and the question arose as to
their attraction relative to that of both cattle and game animals (Boyt and
co-workers, 1978).   An experiment was therefore designed to test the relative
attraction of donkey to tsetse in competition with cattle and game.   As the
experiment was carried out in an area midway between the low Zambezi
valley and the high plateau, the opportunity was taken of checking whether
the overwhelming attraction of cattle in the low veld also applied at higher
altitudes.   The experimental site was in open grassland with fringing woodland infested with G. morsitans and G. pallidipes.   A wide range of game
animals was present including elephant, bushbuck, kudu, reedbuck, impala,
duiker, buffalo, warthog, bushpig, lion, leopard, hyaena, jackal and
baboon.   Nineteen donkeys, 19 donkeys with 19 cattle, and 19 cattle were
grazed in the experimental area for separate weekly periods, no animals
being grazed in the separate intervening weeks.   A similar test with 40
sheep, 40 goats and 20 oxen were carried out at a later date.   Throughout
the test period engorged tsetse were collected daily from resting and refuge
sites and their blood meals tested by precipitin tests.
The results showed that the introduction of donkey resulted in a marked
increase in feeds on equids, and a concurrent decrease in meals from the
natural hosts, suids and bovids.   In the third week, in the absence of
donkeys, the trend was reversed.   The introduction of donkeys and cattle in
the fourth week again increased the proportion of feeds on equids and even
more so on bovids at the expense of suids, the trend being reversed the
following week when cattle and donkeys were recalled.   It was clear from
these experiments that both cattle and donkeys are freely fed upon by
G. morsitans and G. pallidipes in the presence of game.
In the experiment with cattle, goats and sheep the results of blood-meal
identification confirmed the attraction of oxen over goats and sheep, already
established in the low veld.   They showed, moreover, the attraction of cattle

to both *G. morsitans* and *G. pallidipes* in the presence of a wide range of game animals, including other bovids such as buffalo and kudu which are attractive host animals to tsetse.

CHAPTER 6

Horse Flies and Deer Flies (Tabanidae)

i. Introduction

Horse flies and deer flies include some of the largest blood-sucking biting pests of cattle and horses. This, together with their persistence of attack and their daytime activities, makes their presence obvious wherever they occur. Their capacity to take in unusually large blood meals, and to persist in attacking nearby hosts when their feeding is interrupted — as frequently happens because of the painful bites — makes these flies well-suited for the mechanical transmission of certain pathogens, in particular some of the animal trypanosomes. In Africa, their role as vectors of animal trypanosomiasis becomes particularly clear in areas where that disease occurs beyond the limits of distribution of tsetse flies, for example in areas of the Sudan. But even within tsetse areas themselves, their significant role is being increasingly recognized, as for example in the Republic of Zambia where efficient control measures against tsetse do not necessarily affect tabanids.

In the transmission of human and animal disease agents tabanids have been found naturally infected with a wide range of pathogens, and their capacity to transmit many of these has been confirmed experimentally (Krinsky, 1978). Perhaps the longest known direct involvement of tabanids with human disease in particular is the part played by species of *Chrysops* in the transmission of the human filarial infection, loiasis, in the Cameroons and adjacent parts of the West African equatorial rain forest. This association prompted a series of intense studies in that area over a period of about 10 years from the late 'forties onwards. The bulk of our present knowledge about those vector species was gained during that period, since which time the emphasis has been increasingly directed to the parasitology of loiasis, especially the complicated relationship between human loiasis and simian loiasis.

Elsewhere in Africa, the main interest from the point of view of the development of sampling and capture techniques and other study methods, has arisen as a result of investigations originally directed against other established disease vectors such as tsetse flies and mosquitoes. The bulk of recent progress in the study of tabanids, however, is provided by North American work during the last 10-15 years, in which a range of new ideas has been injected into the basic problem of establishing behaviour patterns of tabanids in general.

ii.   Studies on Tabanid Vectors of Loiasis in the Cameroons

To return to Africa, early work in the Cameroons established that the two main vectors of loiasis, *Chrysops silacea* and *C. dimidiata*, were day biting species whose biting activities were normally restricted to the forest canopy; but that under certain circumstances these flies would descend to attack human hosts at ground level (Gordon and co-workers, 1950). In subsequent studies on these biting patterns, several features were disclosed which are of special relevance to the general problems assessed in the present book. One of these lines of investigation was prompted by the observation that a single human, or bait boy, sitting motionless in the forest only received minimum attention from biting flies, but when the subject started moving around — in the way that woodcutters for example would normally do — he was much more pestered by attacks from *Chrysops* (Duke, 1955). Furthermore, if the single static boy was accompanied by a wood fire, large numbers of *Chrysops* began to appear, this attraction being increased nearly five-fold in the case of *C. dimidiata* and over ten-fold in the case of *C. silacea*.

In experiments designed to quantify these observations and experience, two groups of eight trained boys were used who acted as both bait and collectors. Over a period of 5 days, one group remained static throughout the day on the forest floor, while the other group kept moving around, stopping only at intervals. The behaviour of the moving group in this respect was intended to simulate the normal movements of the dominant forest monkey or drill (*Mandrillus*) which moves around in groups on the forest floor by day, and to attract the attention of attacking *Chrysops*. On the last two days of the test, both groups (16 boys) all remained as static bait on the first day, and then all moved around in a body on the second day.

The results of the static bait experiment showed that the group of boys attracted many more *Chrysops silacea* than the single boy bait; the most significant aspect, however, was that the increase in attraction to the larger group was such that the individuals in these groups were attacked at a higher intensity than the single bait. Using the notation of "flies caught per boy per hour" (FBH), this figure for the single isolated bait was 0.25; with the 8-boy bait the figure for each member increased to 0.38, and with the 16-boy group to an FBH of 0.48.

In the case of the moving groups, the intensity of attack during actual movement was lower than with the corresponding static groups of 8 and 16. But when the moving groups stopped for short periods, the intensity increased and exceeded that of the static groups, up to an FBH figure of 0.47 for the group of eight, and 0.64 for the group of sixteen.

Until 1954 all biting studies on these eesentially day-time active species of *Chrysops* had been confined to daylight. The microfilarial stage of the parasitic worm, *Loa loa*, had long been known to show a diurnal periodicity in the peripheral blood vessels of the human loiasis subject, and accordingly transmission by a day-biting insect vector was postulated. But in that year, it was found that the microfilariae of the particular form of *Loa* occurring naturally in drills exhibited a nocturnal periodicity, suggesting strongly that transmission of that form must be by a night-biting or crepuscular tabanid vector. Attention was therefore directed to the after-dark biting activities of *Chrysops*, further incentive to this approach being provided by the disclosure, in the course of Yellow Fever mosquito research in Uganda, that some tabanid species there were very active in the forest after dark. Over a period of two years systematic 24-hour collections of biting *Chrysops* were carried out in the Cameroons programme on a platform built 85 ft up in the branch of a large tree growing at the edge of the forest (Duke, 1958, 1959). Pairs of fly boys working in shifts provided the combination of human bait and collector. It was soon found that in addition to the predominently day-biting species, *C. silacea* and *C. dimidiata*, biting in the canopy, two other species were also abundant at dusk and in the early hours of darkness. The different general patterns of biting of these two groups of *Chrysops* are well illustrated in Figure 6.1, where the figures are based on hourly collections.

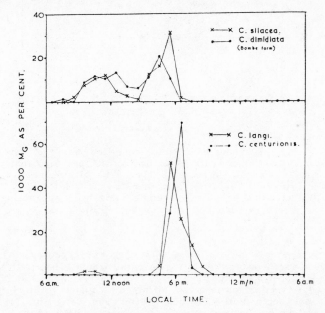

Fig. 6.1 24-hour biting-cycles of the four species of *Chrysops* at canopy level in the rain-forest at Bombe. The hourly biting densities have been computed from the geometric means multiplied by 1,000 (1,000 M_G) of the catches for each hour, and are expressed as percentages of the total 24-hour biting density (After Duke, 1958).

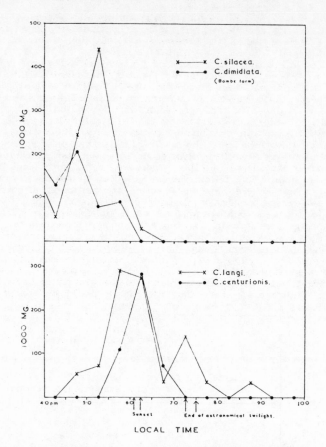

Fig. 6.2 Half-hourly biting densities of the four species of *Chrysops* from 4 pm to 10 pm at canopy level in the rain-forest at Bombe. The biting densities are expressed as the geometric means multipled by 1,000 (1,000 M_G) of the catches for each half-hour period. (After Duke, 1958).

When the collections are broken down into numbers caught every half-hour, a clearer pattern emerges of the biting events around the critical sundown, dusk to darkness period (Figure 6.2). *Chrysops silacea* and *C. dimidiata* show a somewhat similar pattern of biting at canopy level. Their biting activity does not begin until 2-3 hours after sunrise; this is followed by morning and afternoon peaks, after which biting falls sharply to zero at sunset. In marked contrast, the biting habits of the other two species show a crepuscular pattern, with almost all the biting activity taking place between 4 pm and 9 pm; *C. langi* has a biting peak in the hour before sunset, while with *C. centurionis* the peak is in the hour after sunset.

The crepuscular biting habits in the canopy of the last two species would suggest that their normal hosts are likely to be the night-resting monkey population (drills).   The daytime biting *Chrysops* spp. are also attracted to these drills, but at a time when the latter are actively foraging on the forest floor and are much less amenable to any attempts by the hungry flies to bite, much less to take a blood meal.

iii.   Observations on Tabanids in the Course of Yellow Fever Mosquito Studies

One of the most productive sources of new information about tabanid behaviour patterns in Africa has been the classical studies on the trans- mission of Jungle Yellow Fever in the forests of Bwamba in the extreme western part of Uganda from 1942 onwards.   In that work, mainly concerned with mosquitoes, studies on biting activity at different heights from ground to canopy level played a major role.   In the course of those collections on man bait and monkey bait, other biting Diptera were observed and collected. Predominant among these were tabanids of the genus *Chrysops*, species allied to those which had been incriminated as vectors of loiasis in man in the Cameroons of West Africa (Haddow and co-workers, 1950;   Haddow, 1952). In initial studies in the swamp forests of Mongiro, continuous 24-hour catches on four boys were carried out at ground level, and on tree platforms 16 ft, 32 ft and 54 ft in the main canopy.   Later, in the adjacent forest of Mamirimiri, in addition to the platform in the main canopy at 58 ft, an even higher platform at 82 ft was constructed above the main canopy level.   It was at Mamirimiri that observations were first made on a hitherto quite untapped aspect of tabanid behaviour, females of *C. centurionis* being taken predomin- antly at the two highest platform levels.   In addition, the tabanids first became noticeably active at the twilight period.

Continued observations showed that in the wet season *C. centurionis* exhibited a very definite preference for the two highest levels of 58 and 82 ft, the lower of these two apparently being the most favoured level.   While nearly 3000 flies were collected in forest canopy and understorey, less than 50 were taken at ground level.   In the dry season, the vertical distribution is more irregular, with the 22 ft level becoming more important.

While the females were obviously attracted to the human bait at these high platform levels, man was clearly not the preferred host and there is very little man-biting at ground level.   When monkey was provided as bait on the canopy platforms, *C. centurionis* was observed to bite them freely in the presence of man.   Intensified collections from well before dusk until well after dark at 20.30 to 21.00 hours showed that *Chrysops* biting started about dusk,   and reached a peak of intensity about one hour later.   The complete biting cycle worked out later showed that after the post-sunset peak there is very little biting activity throughout the rest of the night or during the day.

The biting habits of *C. centurionis* are in striking contrast to those of its close allies, *Chrysops dimidiata* and *C. silacea*, the main vectors of human loiasis in the equatorial forest regions extending westwards from Zaire.   Both the latter species are mainly arboreal but, in contrast to *C. centurionis*, they are entirely diurnal.   Furthermore, in areas where forest has been cleared, these species will attack man readily at ground level, behaviour which is quite exceptional for *C. centurionis*.   All these observations refer to the blood-feeding females;   nothing was yet known about the normal habits or distribution of the males of these three species.

At a later stage in the Yellow Fever studies a more permanent steel tower was constructed near Entebbe, rising through the canopy to a height of 120 ft (Haddow and Corbet, 1960).   In addition to the routine baited catch, trapping by mercury vapour light was introduced.   The model used permitted a broad belt of light to be cast out sideways in all directions, but prevented it from passing direct upwards.   Insects entering the trap simply dropped

into a deep beaker of alcohol which fitted below the entry cone. The main
outcome of the preliminary trials was that the important factor attracting
tabanids to light was not so much the quality of the light but the precise
location of the trap. The other significant feature was that for the first
time male tabanids were taken in numbers. For routine collections of
tabanids, traps were set at 2 ft 6 ins above ground level. The results
obtained with two dominant species are shown in Figure 6.3. In the case of
one species, *Tabanus par*, both sexes are active throughout the night, and
both show a well-marked peak of activity in the second hour after sunset.
In the case of *Tabanus thoracinus*, the high peak of female activity is
spread over 3 hours, the males on the other hand showing only a slight
activity in the early part of the night, with a major peak shortly before
sunrise.

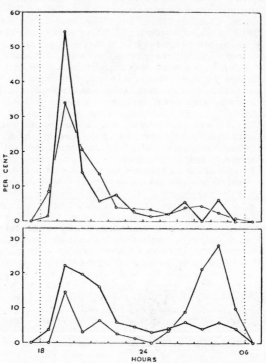

Fig. 6.3 Nocturnal flight patterns of *Tabanus par*
(above) and *T. thoracinus* (below). The figures are
geometric means expressed as percentages to facilitate
comparison. (Thin line = males; thick line = females.)
(After Haddow and Corbet, 1960).

Work on the high forest tower showed that *Chrysops centurionis* was
also taken in light traps and at the highest level above the canopy. Male
C. centurionis were also taken in small numbers sufficient to indicate a
general pattern of activity at different periods of the night and at dawn.

v.   Tabanid Studies in Tsetse Fly Surveys

The Morris trap was originally designed in West Africa for studies on *Glossina palpalis* and *G. tachinoides* (page 84). Later, a modified form was found to be very effective for one of the game tsetse in East Africa, namely *G. pallidipes*. In the course of that work in Uganda some information was gained about East African tabanids attracted to these traps, but it was not until the original trap was used in a tsetse survey in Liberia in West Africa from 1959 to 1960 that opportunities arose for investigating tabanid reaction in more detail (Morris, 1963).

These observations were carried out in the extreme north of Liberia close to the Guinea border but still within the zone of Guinea Rain Forest. The main object of that research programme — tsetse flies and sleeping sickness — dictated the standard capture methods to be used, namely hand-catching with nets by teams of fly collectors, and the mechanical Morris trap. This parallel series of two capture methods was simply extended to include tabanids which were particularly diverse and abundant in the forest region. The traps had the advantage of operating mechanically for 24 hours a day and 7 days a week, so that species with quite different diurnal biting cycles would be equally exposed to the influence of the trap.

Modifications in the design of the basic trap with regard to size and colour were tested on *Glossina*, and the effect of these could also be observed with regard to tabanids. For example, a large size model, 4 ft long and 4 ft high, as compared with the standard 2 x 2 ft design (Figure 6.4), proved to be particularly attractive to *Tabanocella* and some species of *Tabanus*. Beyond that the main objective of the project did not permit further experimentation in trap design to meet the exclusive needs or reactions of tabanids, but nevertheless some general guidance was provided about the relative value of the two sampling methods and about gross seasonal changes in fly abundance. Over 50 traps were eventually in use in this survey, with 6–14 traps covering from $\frac{1}{2}$ to 1 mile of river at each of several widely-spaced localities.

Of the 36 species of tabanid recorded, many were taken in both traps and by hand-catching. Some, including all four species of *Haematopota*, were taken only on man, and about three or four other species, recorded among those taken either regularly or occasionally, were caught only in traps.

Fig. 6.4   A double-sized Morris trap, 4 x 4 ft, used for trapping tabanids in Liberia. (After Morris, 1963).

There seems little doubt that if this investigation could have been continued and devoted exclusively to tabanids, even greater insight would have been gained about the behaviour reactions of these flies to Morris traps as well as to more varied trapping and sampling methods.

Fortunately, although this line of approach to tabanids lapsed at the end of the Liberian tsetse survey, the possibilities of the Morris trap were revived after a few years by American workers with regard to North American species. In Africa itself the reactions of tabanids to models and other attractants was once more taken up in Rhodesia, this time as an adjunct to the intensive work on the reactions of tsetse to models and to improved methods of capture by sticky surfaces and electric grids.

v.   Between 1961 and 1963, over 200 24-hour catches on cattle bait were carried out in Uganda as part of a tsetse research programme based on the N.E. shores of Lake Victoria. Tabanids, as well as other biting flies such as *Stomoxys*, were taken in these tsetse orientated captures, and the continuous records kept have provided a unique insight into biting activities and biting cycles of these flies (Harley, 1965). This work was based on six catching stations, of which three were just inside the moist semi-deciduous forest at the edge of the water, and three in the narrow strip of grassland between that forest and the continuous evergreen thicket extending away from the lake. Black cattle were used as bait, one at each station, alighting flies being caught by hand net. No males were taken, the catch being exclusively females.

The continuous records provided a complete picture of seasonal abundance of several tabanids; for most of these the period of greatest abundance was May to August, extending from the end of the rains until the beginning of the dry season. Depending on rainfall, a second increase in numbers was liable to occur in November and December towards the end of the second rainy season.

The unique value of continuous 24-hour catches of this kind is to provide information about diel biting activity patterns. The records on this aspect extended to 19 species of tabanid and 4 species of *Stomoxys*. The biting patterns of 8 tabanids are shown in Figure 6.5.

With four exceptions, all examples of *Tabanus*, *Ancala*, *Evancala* and *Haematopota* were taken only between sunrise and sunset, and mainly in the middle of the day with peak activity from 11.00 to 14.00 hours. The actual form of the curve tended to show slight variations according to season, with several species exhibiting a slightly earlier peak (at 11.00 to 12.00) in the dry season than in the rainy season (12.00 to 13.00 hours).

In contrast, four species of *Haematopota* exhibited bimodal patterns of biting, with one peak in the early morning a few hours after sunrise, and a second peak an hour after sunset.

The two species of *Chrysops* taken were completely diurnal, with a peak at 14.00 to 15.00 hours, resembling in this aspect the behaviour of *C. dimidiata*, the vector of loiasis in W. Africa, but differing markedly from the nocturnal pattern of *C. centurionis* in Uganda.

vi.   Outdoor Resting Sites of Tabanids in Zambia

Increased knowledge about the normal resting sites of tsetse has drawn attention to the scarcity of information about this aspect of tabanid ecology. The increasing realization of the role of tabanids as mechanical transmitters of trypanosomiasis has provided further incentive to examine this question in Zambia (Oliwelu, 1975, 1977). Over two rainy seasons, which mark the peak

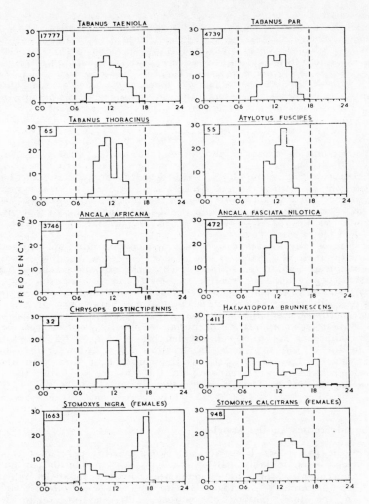

Fig. 6.5 Activity of eight species throughout the 24 hours. The figures are geometric means expressed as percentages. The number of insects on which each diagram is based is given in the top left corner. (After Harley, 1965).

of tabanid abundance, extensive collections were carried out in a variety of bushes, boles, treetops, aardvark holes, branches and fallen logs. Because of the nature of the surface of many of these sites, an indirect method of applying the sticky adhesive adopted (tanglefoot) was used. Boles, branches and fallen logs were banded with paper, while bushes and treetops, as well as aardvark holes, were covered with netting. The outer surfaces of the paper and netting on all sites was lightly treated with tanglefoot, and in all cases the colour of the paper and netting was chosen to blend with the site under observation. Collections were made at intervals throughout the day.

The greatest variety of tabanid species were found in boles and tree-tops, while the fewest species were recorded from branches.   No tabanids were found in the aardvark holes.   Four metres was the maximum height at which resting tabanids were found, and it appears that in the miombo type of woodland in which these observations were made the bulk of resting occurred within 1-2 metres of ground level.   Despite the large number of tabanids collected, females with freshly engorged blood meals were evidently rare, and consequently there was no opportunity of making use of precipitin testing techniques as a means of determining the host preferences of different species.

In the tsetse experiments carried out in Rhodesia with regard to host-finding, opportunities arose of including tabanids in some of these observations (Vale and Phelps, 1974).   The trapping devices were based on the use of electrified nets encircling static bait, tabanids approaching the bait being stunned on contact with the nets and falling into sticky trays below.   In addition, attraction to stationary odour sources in the absence of visual stimuli was studied by the use of live baits hidden in roofed pits.   The attraction to mobile baits was performed in precisely the same way as described for tsetse flies (page 90).

The figures reveal several points of contrast and of similarity when compared with those for tsetse flies under the same conditions, especially *Glossina morsitans* and *G. pallidipes*.   The most abundant tabanid, *Mesomyia decora schoutedeni*, appears similar to tsetse in its upwind approach to stationary baits and in its response to visual and olfactory stimuli from such objects.   As with tsetse, there is no indication that odours greatly increase the attraction to mobile baits.   In contrast to tsetse, the tabanids and pangonids show no marked avoidance of human odour.   Moreover, most *M. decora schoutedeni* approach ox plus man at heights of 1.1 to 2.2 m, as compared with the height of 0.0 to 1.1 m for tsetse.   In addition to the tabanids — mostly females — attracted to bait, smaller numbers of flies were also intercepted and electrocuted when ranging in the absence of bait, and in these samples males as well as females were taken in equal numbers.

vii.   Tabanid Studies in the Americas

Horse flies and deer flies are well known biting pests of both man and domestic stock   — particularly cattle and horses — in the United States and Canada.   Consequently they have long been studied on their own account, and not — as in Africa — incidentally in the course of other biting insect studies.   In the long history of capture and sampling tabanids in the Americas there has been an evolution of techniques which makes an interesting comparison with the not dissimilar evolution of capture methods with tsetse flies.   Methods used in some of the earlier investigations 30 years ago in New York State closely resemble ox-bait catches for tsetse, tabanids being netted off a tethered black Holstein for half an hour between 1.00 and 3.00 hours each day.   Catches were carried out on five days each week, and the figures for the average number of flies caught per hour per week were plotted to depict the seasonal incidence (Tashiro and Schwardt, 1949).

By the late 'sixties several trapping or sampling methods, in particular the Manitoba trap and the Malaise trap, were available, and progress in tabanid studies in the last ten years has been concerned with comparing the performance of these and other techniques under a variety of conditions.   In general the study methods evolved have been quite uninfluenced by work on tabanids in other parts of the world, but there is one noteworthy exception to this.

In an extensive faunal study of the tabanids in Maryland and New

Jersey (1966–1967) attention was concentrated on two different trapping or sampling devices (Thompson, 1969; Thompson and Pechuman, 1970). One of these, the Manitoba fly trap, was already a well-established technique, while the other was a newcomer from Africa in the form of the 'animal' trap or Morris trap (page 84). The Manitoba horse fly trap, first designed and tested in the Canadian province of that name, is essentially a plastic cone resting on a tripod with a collection container fitted at the apex (Figure 6.6). Suspended beneath the cone is the attractant object in the form of a black sphere, beach ball or similar object. Various modifications of the basic design were tested to improve stability, simplicity of operation and ease of transport, as well as prevention of ant invasion by treatment of the tripod legs with tanglefoot. The Morris trap also underwent some modification, but without interfering with the basic design and principle (Thompson and Bregg, 1974).

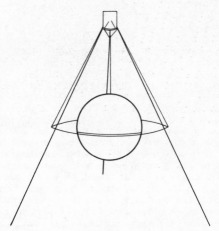

Fig. 6.6   Modified Manitoba horse fly trap.   (After Thompson, 1969).

In the first trials in 1966, one Manitoba trap and one Morris trap were erected in each of three sites 200 m from each other, all in forest openings near water in areas representing different tree associations in the Great Swamp National Wildlife Refuge. In the following year Morris traps were re-sited in fields so that their performance in open situations could be evaluated. The results confirmed previous experience with regard to the high performance of the Manitoba trap as it collected all known species from that area, and proved to be highly selective for tabanids as against other insects. In contrast, the Morris trap proved inadequate for many species, but efficient for others, in particular *Tabanus quinquevittatus*, a widely distributed livestock pest (Figure 6.7). A Morris trap set up in a pen or corral with two horses trapped an approximately equal number of *T. quinque-vittatus* as could be observed directly attacking horses within the same period.

In both Morris trap and Manitoba trap catches were almost entirely female flies. These two techniques have in common the fact that they are purely mechanical and operate throughout all the hours of fly activity. In these preliminary studies these methods were supplemented by a third, namely hand-catching tabanids in flight by means of hand-nets swung over the head. This was found to be particularly useful for deer flies (*Chrysops*) but less so for horseflies (*Tabanus*). It was concluded that there are great

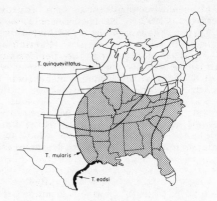

Fig. 6.7 Ranges of *T. quinquevittatus*, *T. mularis* and *T. eadsi* in the United States and Canada. (After Thompson and Pechuman, 1970).

advantages in using a variety of methods to study tabanid flight activity and abundance.

In the evolution of sampling and capture methods for tabanids, work in North America has become more and more concentrated on two basic methods, namely the Malaise trap (Figure 6.8) — with various modifications, and with or without supplementary CO_2 (Anderson and co-workers, 1974; Roberts, 1972, 1977) — and the Manitoba trap and allied forms of decoy. The Malaise trap in its original design and the Manitoba trap are both mechanical, without any bait attractant. The Malaise trap is a non-visible interception trap, operative by day or by night, while the Manitoba and allied decoy traps attract flies visually and are only functional in daylight.

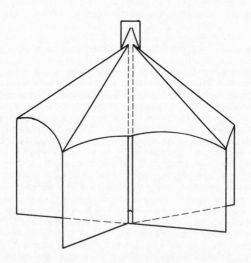

Fig. 6.8   The Malaise flight interceptor trap.

A great deal of ingenuity has gone into improving the design and performance of these traps according to requirements, as well as according to the physical damage and wear-and-tear to which they are likely to be exposed under different climatic conditions. Particularly instructive have been the several investigations comparing the performance of these two main techniques under similar conditions, with or without various modifications. A useful starting point is a simple comparison between Malaise traps, which trap both sexes, and aerial netting or sweeping to collect the female biting flies attracted to bait — usually the human collector himself — carried out in New Jersey (Tollamy and co-workers, 1976). In order to establish some uniformity in capture by aerial sweep net — so often a very variable and subjective method — samples were taken regularly along fixed routes by making ten overhead sweeps with the net, repeated three times, every 15 minutes. Equal catching time was spent in each of the five study areas, and covered the entire daylight period up to and beyond dusk. One Malaise trap was placed in a tabanid flyway in each of the five sites.

The results showed that aerial netting sampled all dominant deer flies (Chrysops) far in excess of those taken in Malaise traps. For example, in one season a total of 7594 Chrysops vittatus were net-caught, compared with only 96 trapped trapped in the Malaise. In contrast, the Malaise trap sampled greater numbers of all dominant species of Tabanus, with one exception.

In another study in Mississippi (Roberts, 1976b) the comparative effect of six trap types, including two types of Malaise trap and two designs of canopy trap (Adkins and co-workers, 1972), five of the six traps were tested in pairs, one with CO_2 and the other without. Although the study showed that different species tended to react to different trap designs in different ways, the results were not sufficiently positive to indicate clear-cut reactions of assistance in defining behaviour patterns.

There are clearly many factors determining the efficacy and performance of Malaise traps, among which colour and age of trap appear to be important (Roberts, 1972, 1975). Quite small differences or changes in the traps are also important in determining how effectively flies directed into this interceptor type of net are directed into the apical collector, or whether they manage to fly back out the way they came in.

The Malaise trap itself has also been used to study vertical flight patterns (Roberts, 1976a), the trap being modified by reducing the size of the opening from the normal 1.2 m to 0.6 m high, then placing the trap on a screened platform preventing tabanids from entering the trap from below. Traps were set up or suspended at five levels from 0.0–0.6 m near ground level, up to 3.6–4.2 m. The results showed that, depending on the species, 72–88% of the flies were trapped at the two lowest elevations. Of these two levels, the majority of tabanids were taken at the 0.9–1.5 m level rather than at the 0.0–0.6 m level. The results gave a clearer picture of the concentration of tabanids in a flyway, where only a low percentage appeared to fly above 1.8 m. There was also little tendency for flies to be attracted above that level, even when positive attraction was provided by supplementing the Malaise trap with CO_2.

In the varying response of different species of tabanid to different designs of trap, or to different modifications of the same trap, many different factors — physiological or environmental — are probably involved. One of these has recently been the subject of special investigation, namely the biting preference shown by different species for particular body area of the host (Mullens and Gerhardt, 1979). It has long been recognized that different species tend to select specific areas of the body when attacking and feeding on livestock, but it is only recently that this question has been re-examined in a more precise and quantitative manner.

Observations were made on cattle which were either tame enough to

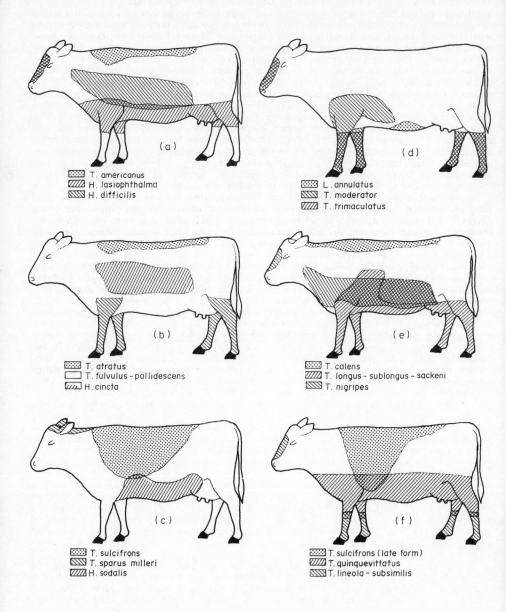

(a)

T. americanus
H. lasiophthalma
H. difficilis

(d)

L. annulatus
T. moderator
T. trimaculatus

(b)

T. atratus
T. fulvulus - pallidescens
H. cincta

(e)

T. calens
T. longus - sublongus - sackeni
T. nigripes

(c)

T. sulcifrons
T. sparus milleri
H. sodalis

(f)

T. sulcifrons (late form)
T. quinquevittatus
T. lineola - subsimilis

Fig. 6.9 Numbered body regions of cow bait, and areas of the body to which different species of tabanid are mainly attracted.   (After Mullens and Gerhardt, 1979).

allow the observer to approach and net flies as they landed, or at least sufficiently tame to allow an observer to stand for long periods at a distance of 10-15 ft, identification of landing species being assisted where necessary with the aid of binoculars.   In the choice of animal host for these experiments, no differences were noted between the attraction of animals of different colours.   The point of each landing during the period of the day from 12 noon to 5 pm was marked on a cow diagram.   In addition, hair depth was measured in 24 body areas in order to obtain a mean hair depth for each region.   Tabanid mouthparts and body lengths were measured for 10 randomly selected females of each species.   Of the total of 44 tabanid species observed, 19 were taken in sufficient numbers to merit analysis.

The results showed that most species have definite preferences for particular body areas (Figure 6.9).   Of the five largest species observed, four landed in significantly high numbers on the upper torso of the animal, while the fifth preferred the upper and middle body.   Several medium species preferred the face and legs, while another group of species, including *Tabanus quinquevittatus*, fed mainly on the lower body.

The influence of hair length was demonstrated by the fact that only the largest tabanid species were observed regularly attacking the upper torso of cattle (Region 4), which has a heavy hair covering.   The smallest tabanids tended to attack areas of thin hair and easily accessible skin, and it appears that the hairy coat is a physical barrier to many species.   Exceptionally, this generalization breaks down in the case of very small flies capable of burrowing between the hairs.   Within these general areas of preference, other factors such as tail action, head swinging, and violent stamping and kicking tended to concentrate flies in the thoracic part of the body rather than the hind quarters.

The characteristic feeding patterns of various tabanid species probably have an influence on their amenability to different collecting methods.   Traps requiring entry from the bottom such as the Manitoba trap, are poor collectors of high-feeding species.   In this connection the success of the Morris or 'animal' trap with some species, such as *Tabanus quinquevittatus*, but not others (page 129) may well be due in part to the low body region feeding preference of that species (see Figure 6.9).

CHAPTER 7

Blackflies (Simuliidae)

i) Introduction. Special ecological requirements of
blackflies. Breeding habitats in running water.
Simulium control.
ii) Studies on pest and other blackflies in Canada and the
U.S. Bird-feeding habits of ornithophilic species.
Studies in Algonquin Park and in Wisconsin. Man and
livestock pest *Simulium*. Mark-release-recapture
experiments in Canada.
iii) *Simulium* studies in W. Africa. *S. damnosum* and onchocerciasis. Man-biting rates. Mark-release-recapture
experiments in Ghana and the Cameroons. Special
problems on the OCP (Onchocerciasis Control Programme).
Flight range and re-invasion. The search for outdoor
resting sites. Light trapping and other recent techniques. Studies on visual and olfactory stimuli in
the Cameroons. Host selection studies in Central
America. Bird-feeding work in W. Africa and the
Sudan. Behaviour patterns of members of the *Simulium
damnosum* complex.
iv) Truck trap studies in England and the Sudan.
v) Australian *Simulium* studies. Tent traps and sticky
traps. Sweep netting for outdoor populations.

i.   Introduction

Blackflies are among the most important and widespread of the small
biting Diptera. Many of the problems in studying the adult behaviour and
ecology are essentially the same as for other groups of small, predominantly
day-biting flies, and centre on such basic activities as biting behaviour and
biting cycles, host selection, flight paths, flight range, longevity, etc.
However, there are certain unusual characteristics of the Simuliidae which
have influenced or determined the course of research on adult behaviour
patterns. First is the fact that while species of *Simulium* thrive under a
wide range of climatic conditions from tropical to temperate — even subarctic
— the larval stages are always dependent on running water habitats. Some
species prefer small streams or brooks, others are adapted to larger streams
or small rivers, while still others are associated with the most turbulent
rapids and fast-flowing stretches of such great rivers as the Congo, the
Niger and the Volta. Across that wide range of larval habitat, the common
factor of complete dependence on running water remains paramount.

These characteristic larval and pupal habitats have long attracted the attention of a long succession of entomologists and biologists. Studies on larval ecology have been assisted by the fact that the identification of *Simulium* species in the aquatic stage can in many cases be clearly determined by the form and branching pattern of the pupal spiracles, a relatively simple diagnostic feature. Consequently a great deal can be learnt about the distribution of different species and their specific breeding requirements on the basis of running water studies and surveys alone.

A second characteristic is that the complete restriction of immature stages to running water habitats, often very clearly defined types of running water, has made the larval stages a much more obvious target for control measures than the widely diffused adults. This aspect of research has received impetus and encouragement since the appearance of DDT and allied synthetic organic chemicals with unique larvicidal properties made *Simulium* control feasible. Chemical control of *Simulium* larvae with more recent organophosphorus larvicides dominates the whole *Simulium* field at the moment, and is the motivation behind one of the world's greatest insect vector control programmes, namely the World Health Organization coordinated Onchocerciasis Control Programme (OCP) centred in the Volta River basin of West Africa (Davies and co-workers, 1978; WHO, 1977, 1978).

Undoubtedly, the effect of this emphasis on larval surveys and larval control has been to reduce research activity on adult behaviour patterns disproportionately to other fields of biting flies. This trend has also been further accentuated in many areas by the real difficulties facing competent entomologists in trying to unravel even the outlines of behaviour patterns regarding adult flight and feeding habits. However, there is now evidence of a swing in the pendulum. The intensification of control measures has postulated the need for more critical methods of evaluating the effects of such measures on the adult blackfly population. The problem of re-invasion of controlled areas, and of pinpointing the source of such outbreaks, has demanded a new and vigorous approach to flight range and flight patterns, and in particular the physiological composition and age structure of such invading populations.

Simulium adults have been studied intensively not only in their role as troublesome pests of man and more serious biting pests of livestock, but also as vectors of ochocerciasis, the filarial infection of man which in areas of intensive transmission may lead to serious ocular lesions and even blindness. The long record of investigations in Canada and parts of the United States exemplifies studies directed against *Simulium* as a pest biting fly, causing serious livestock losses over wide regions, and interfering with man's activities and land development in others. Much of the North American work has also been concerned with *Simulium* species which are mainly bird-feeders, some of which are vectors of parasitic diseases of domestic ducks, fowls, geese and turkeys, of considerable economic importance. The Canadian work in particular, therefore, covers a wide sphere of interest, and has been marked by outstanding contributions to our knowledge of *Simulium* in general. Somewhat similar problems of *Simulium* as a man-biting pest, and dangerous pest of livestock, occur in many other temperate zones, particularly in parts of Russia and Central Europe.

In contrast to such situations are those cases where *Simulium* has been investigated mainly in its role as vector of ochocerciasis of man, particularly in tropical West and Central Africa as well as several countries of Central America such as Guatemala, Colombia and Venezuela. In the last 5-6 years the approach to such *Simulium* vector studies has been dominated by the WHO Onchocerciasis Control Programme in West Africa, which has stimulated rapid progress in some aspects of *Simulium* ecology and re-defined the importance of the many problems still remaining.

It is naturally impossible within the confines of this chapter to do justice to all the noteworthy work in this rapidly increasing field of research in so many different countries. It is considered that perhaps the best approach to this difficult study is to concentrate on the problems encountered in two main areas of intense research, namely Canada and Africa, and to incorporate in that review as many significant developments as possible from other countries such as the UK, Australia and the USA, where notable contributions have been made in the area of field study methods relevant to *Simulium* problems in general.

ii.   Studies on Pest and Other Blackflies in Canada and the US

Simulium work in Canada, and adjoining states of the US, has been concerned with three main problems: the ornithophilic or bird-biting species responsible for avian disease;   man-biting species and pests of livestock; and studies on general problems of flight by means of mark-release-recapture techniques.

One of the major contributions to the study of bird-biting species of *Simulium* was carried out in the wildlife reserve of Algonquin Park, Ontario, and was concerned with the feeding preferences of species known to be preferential bird feeders, rarely recovered from mammals (Bennet, 1960; Fallis, 1964).   About 20 species of bird were exposed in four different habitats in the Park, two associated with the lake and lake shore, and two with the forest about ½ mile distant.   The birds were confined in cages, either 8 inches or 12 inches cube according to host size.   Most of the exposed birds had hoods put over their heads to keep them relatively motionless. The smaller birds were restricted in movement by means of chicken mesh tubes.   The bird cages were placed on two-foot squares of white plywood, and after exposure they were covered with a 2 x 2 x 2 ft nylon screen collecting cage with an open bottom.   Flies trapped inside were collected through a sleeve by means of an aspirator.   Bait cages were used to study activity at different heights.

The results showed that some species, such as *Simulium rugglesi*, were usually collected at ground level at the lake shore, while others were generally confined to the forest habitat, where the majority feed on birds several feet above ground level.   By offering a choice of bird hosts it was shown that *S. rugglesi*, which is confined to a narrow zone at the lake edge, had an overwhelming preference for ducks over crows, grouse, sparrows and others.   Furthermore, *S. rugglesi* exhibited marked preferences within the duck family itself.

Further studies on bird-biting species in the adjoining American state of Wisconsin are particularly instructive in illustrating the development and improvement of more efficient study methods for trapping and sampling *Simulium* attracted to bird bait (Anderson and de Foliart, 1961).   This work was carried out in areas which had been developed for attracting and holding breeding populations of migratory waterfowl.   One of the study areas also supported several flocks of wild turkeys.   About a dozen different species of bird, including both wild and domestic, were used in bait experiments, as well as mammals such as squirrel, skunk, cottontail rabbit, domestic rabbit and deer.

With some of the larger birds such as turkeys and cranes, as well as larger mammals, adult *Simulium* which had settled on the host were collected by means of suction tube or aspirator.   In most cases, however, the avian and mammal hosts were exposed in open mesh cages through which *Simulium* nad access.   The actual method of trapping attracted flies consisted of exposing the cages for short periods of 30 or 15 minutes, at the end of which period a cardboard trap in the form of a box slightly larger than the cage was placed over the host cage.   In its first form the trap box had windows, covered with cheesecloth, cut in the sides and top.   When the trap box was placed over the host cage after 30 minutes exposure, trapped flies flew to the window.   The box was then lifted free and the open bottom quickly covered with a piece of cheesecloth attached to one side of the box.   Flies were collected by aspirator through sleeves at either end of the trap.

In its later form the trap was used as a 'blackout' box, the only source of light being a 5-inch square hole in one of the upper corners.   The actual trap part of the device was a 6-inch cheesecloth-covered cage set over the opening on top of the box.   After exposure of the bird or mammal bait for a 15-minute period, the cage was covered by a blackout box, and a sliding panel between it and the trap above was partly opened to allow flies attracted to the light to pass into the top cheesecloth cage component.   This trap with its contents could then be removed and replaced with an empty one while the hosts were exposed for a further period of 15 minutes.

Trap cages of this kind were also used at different heights above the ground, using a dual pulley system to lift both the host cage and its blackout trap box.   This trapping system was found to be an efficient method for comparing the attraction of simultaneously exposed hosts.   Both fed and unfed *Simulium* readily moved into the top trap when the blackout box was placed over the host cage.   As many as 300-700 flies were caught per host per exposure in the larger catches, and this represents about 90% of the total flies covered by the blackout box.   As the method allowed a host to be exposed for two 15-minute periods each hour, peak periods of feeding activity could be accurately determined and correlated with daily meteorological data.

Host selection was studied by laying out an experimental design of trapping, the arrangement being rotated from time to time in order to eliminate bias.   Seven species of blackfly females were collected from the bird hosts, none of which were observed feeding on the mammals.   Although most bird feeders fed on all avian hosts exposed, various species were more attracted to birds of certain orders than others.   However, in other cases it appeared that attraction was influenced more by size and surface area of the host than by other factors.

In the marsh habitat, as distinct from the forest habitat, the dominant species was   *Simulium rugglesi*, incriminated as the principal vector of *Leucocytozoon* parasite of ducks in Canada (Anderson and co-workers, 1962). These investigations provided new insight into the feeding habits of *Simulium* in relation to wild ducks and other avian hosts.   S. *rugglesi* is most abundant at ground level in a narrow zone along the lake shore, and density falls off rapidly away from this area, ducks at 100 ft from the shore edge, for example, attracting much lower counts than those 20 ft from the shoreline.

The most common simuliid feeding on turkeys and chickens in Wisconsin is *Simulium meridionale*.   Host preference studies showed that this species had a marked preference for gallinaceous birds over ducks.   Comparative tests with turkey and pheasant-baited traps at ground level and at 20 ft showed a significantly greater number of flies attracted to and feeding on birds exposed in the lower canopy than on similar hosts at ground level.

The increasing use of CO_2 as an attractant for biting flies has also extended to the study of bird-biting *Simulium*.   In work on *Simulium slossanae*, a primary vector of *Leucocytozoon* of turkeys in South Carolina (Moore and Noblet, 1974), the standard capture trap consisted of a 2 x 4 x 84 inch stake placed approximately 2 ft into the ground, with an inverted 50 lb lard can attached to the top of the stake (Baldwin and Gross, 1972).   The can is painted a dark shiny blue — attractive to blackflies — and has a strip of polythene attached to the outside coated with a slow-drying adhesive. The CO_2 attractant was emitted from a 50 lb block of dry ice maintained in a sealed styrofoam cooler.   The sublimating CO_2 is conveyed to the inside of the can through a rubber tube.   Ice blocks are replaced twice a week, and the sticky strip at 7 or 14 day intervals depending on the blackfly collected. These traps were also found to be highly attractive to tabanid flies.

Canadian work on the attraction of *Simulium* to bird bait has disclosed an unusual degree of specific preference on the part of one species. This is *Simulium euryadminiculum*, whose biting habits are completely dominated by its attraction to the common loon (*Gavia immer*). The existence of a powerful olfactory stimulus was demonstrated by the strong attraction of flies even to dead birds, to particular parts of the body, and to ether extracts from the loon's tail. The active source of this unique attractant was eventually found to be located in the uropygial gland. The powerful response of *Simulium euryadminiculum* to ether extracts of this gland has enabled an assessment to be made of the responses of females of this species to various olfactory and visual stimulu (Bennet and co-workers, 1972).

The first point to be determined was the attraction to the extract with and without supplementary CO_2. The second was to determine the response of the flies to (i) loon-like decoys, and to specific areas of these decoys, (ii) 2- and 3-dimensional targets of different shapes and colours, In the first experiment one fan trap (page 142) baited with gland extract, one with CO_2 alone, and one with extract plus CO_2, were all operated simultaneously. In contrast to its effect on other Simuliidae, CO_2 alone is a poor attractant to *S. euryadminiculum*. The combination of both extract and CO_2, however, attracted nearly three times as many flies as the extract alone, indicating some synergistic action.

In the experiments with visual stimuli, all the models were painted with clear tanglefoot, on which attracted flies were held. Preliminary observations had shown that a mesh packet of scent bait lying flat on the ground was less attractive than one suspended in the air, indicating that visual and possibly three-dimensional factors entered into the attraction. Low floating styrofoam decoys representing black ducks were used as models resembling the loon. Gland extract was placed on the backs of the decoys, which were coated with tanglefoot and floated 3-5 feet from the shore. Decoys could be bobbing (i.e. free-floating) or stationary, but this was not found to be a significant factor. Of the total flies collected more than 70% were captured on head, neck and beak, which comprise about 20% of the total surface area. Increasing the length of the model's neck by 6 inches from normal increased attraction, perhaps by presenting a greater landing area.

Experiments with 'head-neck' decoys of variously coloured art bristol board showed that the attractive areas were at the angle of the head and neck, and at the prominent terminal portion of the model (Figure 7.1). In these and similar experiments it appeared that flies selected the most prominent portion of the models on which to land, and that the pattern of fly distribution was further influenced by the exact location of the scent bait on the model. Different colours and colour combinations were also tested on three-dimensional models and showed a gradation of attraction from black, which was most attractive, to white, which was relatively unattractive.

Many of these findings have been incorporated in later mark-release-recapture experiments, described on page 146 (Bennet and Fallis, 1971).

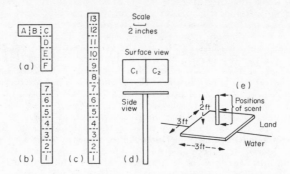

Fig. 7.1   Equipment   and   models   used   in   experiments.
a,b and c to   scale;   d   and e diagrammatic.   (a)
Head-neck model;   white   collars   placed   at   zones   B   and
E in some tests.   (b)   Short   cylinder   model;   white
collars   placed   at   zone   4   or   6   in   some   tests.   (c)   Long
cylinder model.   (d)   Arrangements   used   for   testing
colour attraction of   two-dimensional   models.   (e)   Rotating board on   which   all   trests   with   models   were
carried out.   (After   Bennet   and   co-workers,   1972).

Man and livestock pest *Simulium.*   *Simulium* have long been important
pests of cattle and man in Canada, and they continue to play an increasingly important part in northern development schemes.   The major pests of
man are *Simulium venustum* and *Prosimulium hirtipes,* while *S. arcticum* is a
serious pest of livestock in the prairies (Fredeen, 1958, 1969, 1977).

The long-established studies on *Simulium* biology and control in Canada
and adjoining parts of the United States have built up over the years an
unusually rich foundation of knowledge which continues to exert a strong
influence on *Simulium* research in many other parts of the world.   In the
course of all this work, a wide range of field study methods have been
devised and tested experimentally.   Among the first sampling methods to be
established for adult flies, particularly for man-biting species such as *S.
venustum,* was the standard landing rate count (Peterson and Wolfe, 1958).
A 12-inch square of dark blue serge cloth, divided by white lines into nine
squares, is placed in the lap of the observer;   after a two-minute waiting
period the number of flies landing during the next minute is counted.   When
possible three simultaneous and three consecutive counts are made at two-
minute intervals.   The use of dark blue cloth was indicated by previous
studies on factors attracting biting flies to man.

In a later investigation on *S. venustum* on the north shore of the St
Lawrence River, Quebec, in 1945-56, this basic technique was slightly modified (Wolfe and Peterson, 1960).   An 18-inch square of blue serge cloth was
used, and the landing rate of blackfly was recorded per minute.   This
method was supplemented by two others, namely the biting rate on the fore-
arm of the human bait/collector, and the number of flies collected in a 12-
inch diameter net in ten figure-of-eight sweeps round each collector.   For
the landing and biting rates, as well as sweep counts, three observers were
positioned in a triangle, and they changed places every eight hours.

An example of the results of such techniques is shown in a 24-hour
survey (Figure 7.2).   These reveal evening and morning peaks of activity,

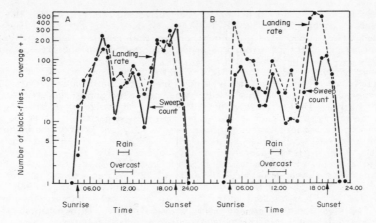

Fig. 7.2 Sweep counts and landing rates of blackflies at hourly intervals during two simultaneous 24-hour surveys in the Baie Comeau area, Quebec, July 1-2, 1955. (A) In the forest near Brisson Creek; (B) in a treeless area at Lac la Loutre. (After Wolfe and Peterson, 1960).

as judged by sweep counts and landing rates. Biting rates were at a much lower level than landing rates, averaging 2.6 per minute during the daylight hours, but they showed similar increases in the morning and evening (Figure 7.3). Further studies showed that the actual timing of the two peaks dif-

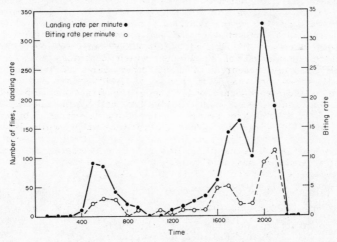

Fig. 7.3 Activity of blackflies at Brisson Creek, near Baie Comeau, Que., during a 24-hour survey on June 23/24, 1956, showing mean landing rates on 9 x 9 in blue cloths and mean biting rates on bare forearms recorded by three observers. (After Peterson and Wolfe, 1958).

fered according to the location of the collecting site, i.e. in the open or under forest cover. Activity in the open started earlier in the morning and continued after sunset for an hour longer than in the forest (Figure 7.2). The depressing effect of wind was also more marked in the treeless collecting site.

More recently, entirely different methods of sampling S. venustum have been tried out in the now famous study area, Algonquin Park, Ontario (Fallis and co-workers, 1967). This study was not only concerned with developing efficient capture methods for adult blackflies, but also in examining various visual, olfactory and other stimuli determining attraction. Over a period of two summers collections were made in fan traps, on 'sticky mannequins' and on flesh-coloured cylinders beside which CO_2 was released in varying amounts. The small tubular fan traps, operated by a 6-volt power supply, measured 4 inches in diameter, by 5 inches long, with four 3-inch blades. The traps were suspended from the ends of 12 ft arms extending from a turn-table, which was moved through $90°$ every 2 minutes during each collection (Figure 7.4). Flies sucked into the trap were collected in attached gauze bags. CO_2 was released beside some of the traps. In some collections heat was provided by hot water bags suspended beside the traps.

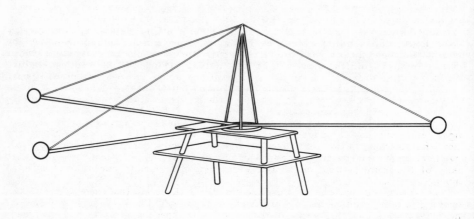

Fig. 7.4 Use of fan traps for collecting blackflies, suspended at the ends of 12-foot arms attached to a turntable placed in a clearing in the woods. Also construction and use of silhouettes for trapping. (After Fallis and co-workers, 1967).

This technique trapped flies that came within range of the air currents generated by the fan, but provided no proof that the flies caught in the trap would have landed. Data on the actual landing aspect was provided by the mannequins and cylinders, each of approximately 1 square yard. Heat and CO_2 were used as stimuli with the cylinders, but only CO_2 with the mannequin. The cylinders, 52 inches long and 9 inches in diameter, were covered with black paper over which a flesh-coloured wax paper was placed and coated with sticky tanglefoot. Cylinders could be set up in either vertical or horizontal positions.

Several collections of flies were made at different times and under different conditions at the end of the 12 ft arms. Some traps had CO_2 released, while others with no CO_2 served as controls.

The results showed that blackflies (over 90% of which were S. venustum) were taken in greater numbers in traps with CO_2 than in controls, and that more flies were captured with increasingly larger amounts of CO_2 (up to 800 ml per minute).

Simulium arcticum is the most important livestock pest in Canada, particularly in the western prairie and agricultural states (Fredeen, 1958). It is notorious for appearing in massive swarms in the early summer, often detertmined by unpredictable factors such as mass emergence from breeding grounds or by changes in wind direction (Fredeen, 1969; Alverson and Noblet, 1976). It is only comparatively rarely a man-biting pest, and is not therefore amenable to standard methods using human bait. There is some evidence to show that the first summer generation of females may develop their ovaries autogenously (i.e. without the normally necessary blood meal) in the first ovarian cycle, reverting to a normal blood-feeding pattern in subsequent cycles. Because of this, the population of emerging females may build up over several days until climatic or other conditions are suitable for the first oviposition. Subsequent to this the now massive population of females will seek their first blood meal as soon as conditions are favourable, and may produce peak biting all within a few hours or less (Fredeen, 1963).

Because of the unpredictable nature and brief duration of many of these attacks of biting on cattle, information about the composition of the blackfly swarms was unobtainable by any of the normal techniques. However, with the cooperation of the livestock owners on the spot netting swarms of attacking flies, over 46,000 flies were caught. In this way it was possible to show that although two or three different species were sometimes present together, most swarms were composed of at least 90% S. arcticum and that many of the more severe outbreaks were almost certainly due to this species alone.

In order to provide a more scientific and acceptable sampling method for longer term studies on S. arcticum a trap was finally devised based on the observation that attacking females of this species seeking a blood meal usually assemble near animals or other objects and from there they usually attack the animal on its ventral surface. In a dark enclosure flies move readily towards the light. Embodying all these principles, a trap was designed consisting of a four-legged frame with the upper two-thirds and the top covered with dark brown plywood or dark blue denim cloth — colours already known to be attractive to certain species of blackfly (Fredeen, 1961). The underside was left open, but direct daylight or sunlight was admitted to the interior only through an opening in the top, over which was placed a clear plastic cone terminating in a removable glass jar. The operating principle was that flies attracted to the model would enter from the underside into the dark interior, and thence to the sunlit apex.

Three types of trap were used (Figure 7.5). The largest or 'cow' silhouette trap, named because of size and shape, was 4 ft high, 5 ft long and 2 ft wide, with an underside opening of 10 sq. ft. The second, or 'sheep' silhouette was 18 inches high, 2 ft long and 1 ft wide with an

Fig. 7.5 (1) Cow silhouette trap. (2) Sheep silhouette trap. (3) Pyramid silhouette trap. (4) Pyramid silhouette trap, collapsed for transportation.
(After Fredeen, 1961).

opening of 2 sq. ft. The third or 'pyramid' silhouette trap was 4.5 ft
nigh, 3 ft square at the bottom of the canopy, with an opening of 9 sq. ft.
These three were tested simultaneously, and for comparison a standard light
trap was operated in the vicinity — but out of sight — from 6 pm to 6 am.
The trapping tests were carried out during the months of major blackfly
activity and showed that, of the three silhouette traps, the 'cow' was the
most effective, this efficiency evidently being a direct function of surface
area and size of opening, regardless of shape.

Rather unexpectedly, in view of the known daytime activity of *S.*
arcticum, the light trap collections exceeded those of the silhouettes, and
even attracted blackflies during the daylight hours when neither the light
nor the suction fan was operating. This may have been due to daytime-
active flies being attracted to the canopy over the light trap where they
would be overcome by cyanide fumes. A further complication was that some
of the largest collections in the light trap were obtained on days when
livestock were not severely affected.

In the context of sampling blood-sucking flies in general, there is a
strikingly close resemblance, both in principle and operation, between these
animal silhouette traps specially designed for *Simulium*, and the 'Morris' or
animal trap designed for tsetse in West Africa (page 84) and later used for
tabanids as well in both East and West Africa (page 125). It is equally
clear that this common approach to a basic sampling problem was arrived at
quite independently in these two different fields, and in mutual unawareness.

The second significant outcome of these studies on *S. arcticum* is the
inclusion of a conventional light trap in the experiment, and its quite
unexpected success. In general, blackfly workers in Canada have long dis-
counted light traps as quite inappropriate to sampling essentially daytime

active insects, even though as far back as the early 'sixties the same technique had met with remarkable success when first tried out in Scotland under
conditions rather similar, and against all logic for flies not supposedly
active after dark. The 'success' of the light trap does, however, raise all
the old problems and misgivings about what exactly is being sampled, and
the extent of the range of attraction of such traps. Are the high catches
illusory in that flies are being concentrated or aggregated from a much wider
area than the limited range of daytime silhouettes, or is it possible that
light traps do not so much sample normal flight activity as to activate
insects normally dormant after dusk? These are all points which have a
bearing on the general problem of determining behaviour patterns of blood-
sucking flies, and they will be reconsidered more fully in the final summing
up.

Mark-release-recapture. Canadian workers early recognized the need
for more precise information about flight range of *Simulium* and dispersal
patterns from their breeding areas. The existence of swarms of pest blackflies long distances from the nearest possible river habitats indicated that in
some cases swarms originated from distances of over 20 miles. In other
cases and with other species, dispersal appeared to be more limited. In
examining this problem over the last 30 years, three distinct methods have
been used.
In the first of the large scale trials, in Saskatchewan, the method used
was the radioactive tagging of larvae in such a way as to be detectable in
the adults subsequently produced (Fredeen and co-workers, 1953). A large
number of full-grown larvae of five species of *Simulium* were marked with
sufficient P^{32} to be readily detected with a Geiger counter throughout the
life of the insect. The marking was accomplished by transferring the larvae
from the normal stream habitat to an aquarium or tub containing a small
volume of the stream water to which P^{32} had been added. After 24 hours
exposure to the radioactive medium they were transferred to a non-radioactive
medium or into a stream for the completion of their development. The tagging
of a large number of blackflies was feasible only in the last instar larval
stage as collection and activation of adults in large numbers was not
practicable.
Although tagged adults detectable by Geiger counter were produced from
the larvae under laboratory conditions, there was complete failure to recover
more than a single tagged adult in nature, from an estimated 700,000 radioactive larvae released. Those workers considered that this failure was not a
condemnation of the method itself, or its real possibilities, but rather a
failure to use recapture methods of sufficient efficiency, and on a large
enough scale, to cover all possible avenues of dispersal. The capture methods
actually practised were regular collections of flies from livestock by means
of nets, together with net-sweeping vegetation along river banks, supplemented by the use of light traps.
It was not until several years later that the real potential of this
larval tagging technique was fully exploited, the two important factors
contributing to success being the improved methods for accumulating very
high larval populations from natural habitats for radioactive tagging, and
the increased efficiency of traps used in a more extensive network of
trapping stations. These trials were carried out in the Chalk River area of
Ontario (Baldwin and co-workers, 1975). In preliminary tests an estimated
260,000 larvae from a very productive stretch of rapids were treated with P^{32}
and released. In the subsequent adult capture programme a total of 261,600
adults were collected, of which 166 were radioactive.
In a later and more extensive series, an estimated 18 million larvae,
representing three species of *Simulium*, were tagged and released to develop
into adults. The efficient larval accumulation technique consisted of
attaching groups of strands from a mop head on an 8 ft rope leader, which

was then weighted at each end and immersed in the stream. Larvae moving downstream congregated on these artificial substrates in enormous numbers and could be readily transferred to the treatment tubs for labelling, and later returned to the river.

The adult capture technique employed the principle of CO_2 attractant and sticky surfaces (Baldwin and Gross, 1972). Large empty lard cans of 50 lb capacity were suspended in an inverted position from branches of trees 4 ft from the ground. A wide strip of plastic covered with a thin layer of extremely sticky substance called 'Roost-no-more' was fastened around the tins. A $2\frac{1}{2}$ lb piece of dry ice, which gave off CO_2 for a period of 12 hours, was wrapped in paper towelling and placed on half the traps each week as attractant. Flies attracted and held by the sticky trap were taken to the laboratory, where radioactive individuals were detected by means of X-ray film. A total of 225 traps were deployed over 900 square miles, the traps being visited weekly in the high-output season from June to mid-August.

The results pertaining to *S. venustum* were particularly interesting. Two days after tagging, a few labelled individuals were recovered at 22 miles in a trap at the extreme limit of the test area. However, the majority of the flies did not appear to move so extensively, high numbers only being recorded up to 7-8 miles. Of the 4675 radioactive flies recovered, the average dispersal from the release point was between 5.8 and 8.2 miles.

In the second example, carried out in Algonquin Park, Ontario, nearly ten years after the original Saskatchewan trials, radioactive tagging was the method of choice but in an entirely different set of circumstances (Bennet, 1963). The test species chosen for these trials was *Simulium rugglesi*, the vector of blood parasites of ducks. This species is found in a narrow littoral zone about lakes, ponds, rivers and streams in the Park, where they feed extensively on anseriform birds.

Populations of wild flies trapped by methods previously developed for this species (see page 137) were allowed to feed on domestic ducks which had previously been injected with radioactive phosphorus (P^{32}) at concentrations which had previously been determined experimentally as sufficient to label the biting flies effectively. After release of fed females, regular collections were made in several areas by capturing wild flies on non-radioactive duck bait. Under these conditions the recovery of labelled flies was encouragingly high, namely 503 (or 13.4%) of 3747 tagged flies released. The high recovery rate, together with the fact that tagged flies were recovered every day subsequent to release up to the 12th day, permitted some conclusions to be drawn on the longevity of the females and their frequency of feeding, as well as their movements and distribution. The actual flight range of *S. rugglesi* was determined by collecting labelled flies at various points in the experimental area. Some individuals were recovered at a capture station 2 miles from the marking site in a direct line but 6 miles by water, this distance being covered in 5-12 days. Marked flies moved both with and against prevailing winds and water currents.

Basically the same technique was used in mark-release-recapture studies on *Simulium slossonae*, a primary vector of *Leucocytozoon* of turkeys in South Carolina (Moore and Noblet, 1974). In this case turkeys were labelled by means of intraperitoneal injection of radioactive phosphoric acid. The recovery technique differed in that, instead of being based on live bird-baited traps, CO_2 baited lard cans were used, these being arranged in concentric circles at distances of 0.5 to 8 miles from the tagging centre. During a three-month study 206 labelled specimens were recovered out of about 56,000 flies caught.

The third method used in Canadian work in radioactive tagging differed from the foregoing in that captured adults, not larvae, were directly tagged for release (Bennet and Fallis, 1971). This was only feasible because the bird-feeding species being investigated, *Simulium euryadminiculum*, is specifically attracted to the common loon (*Gavia immer*) and to ether extracts of the uropygial gland of that bird. The large number of wild-caught flies necessary for the success of such tests was obtained by fan traps provided with the attractant ether extract, flies being collected in mesh bags at the bottom of the trap.

About 15,000 to 20,000 of these captured flies were placed in a cage at one time and at the end of a period of 1-3 hours — sufficient to cause partial dehydration — they were given the opportunity of drinking a solution containing enough radio-isotope to ensure that about 80% of the flies were labelled after 30 minutes. Decay and loss of isotope on marked flies was checked in the laboratory with caged samples tested daily.

Over a period of two weeks approximately 94,000 flies were captured and labelled around the shores of a lake, and then released at one point. Subsequently 127,000 flies were captured from various areas, and of these 204 (1.6%) were found to be radioactive. Most of the marked flies were recovered within 1.5 miles from the point of release, the maximum point of recovery being 5 miles.

iii.   *Simulium* Studies in Africa

The primary subject of *Simulium* research in tropical Africa has been *Simulium damnosum*, the long established main vector of human ochocerciasis in West and Central Africa. Another important vector of much more limited distribution in Central Africa is *Simulium neavei*, the subject of a successful eradication programme in the 'fifties.

Simulium damnosum is a species particularly associated with some of the largest rivers and river systems in West Africa, such as the Niger, the Volta and the Bandama, as well as the middle reaches of the Nile. In general the most productive breeding grounds are in the savannah zone stretching from west to east between the rain forests to the south and the desert to the north.

In the last few years the scope and direction of research on *Simulium damnosum* has undergone a drastic reappraisal. This is due to two dominant and closely related events. The first was the initiation in 1974 of the World Health Organization Onchocerciasis Control Programme (OCP) based initially on the Volta River basin and embracing seven adjoining countries of West Africa (Davies and co-workers, 1978; WHO, 1977, 1978). The second event was the disclosure that *Simulium damnosum*, for several years regarded as a fairly uniform species — possibly including at most a 'forest' type and a 'savannah' type — was in fact a complex composed of at least nine different species, each with characteristic distribution and ecological requirements. The extent of the complex now defined, and still in process of further clarification, marks an important milestone in progress. At the same time this discovery presents a whole new range of problems, not least of which is the reassessment of all previous information pertaining to '*Simulium damnosum*' in its wide sense. Recently gained knowledge about which member or members of the complex exist in each locality may help to reconstruct the exact identity of the '*Simulium damnosum*' originally investigated in those areas. The implications and repercussions of these recent advances will be discussed in more detail later. But for the present review of field study methods it will be sufficient to refer simply to *Simulium damnosum* wherever that has been the object of study in published reports.

As a major vector of human disease, the man-biting activities of *Simulium damnosum* have long attracted considerable attention. The simple and straightfoward collection of biting female flies from human bait has been in marked contrast to the great difficulties in collecting adult flies in any other way. Consequently the man-biting aspect of behaviour has come to dominate field studies on adults of this species, and for a long time now has been the sole standard method for sampling adult populations (Le Berre, 1966; Service, 1977; Thompson, 1976b). Both French and British *Simulium* workers have used the man-biting rate to build up a clearer picture of biting activity in different regions, at different seasons, at various distances from breeding grounds, and also at different periods throughout the day from dawn to dusk. In addition, samples of biting females have been divided into two categories (parous and non-parous) or into three (non-parous, early parous and late parous). Those studies have demonstrated that the biting cycle varies according to those physiological conditions, and that the peak biting of parous and non-parous may occur at different times of the day. A great deal of information has also been built up about which regions of the body or lower limbs are preferred biting sites for the flies. This method covers wide ranges of biting density; at one extreme, in some savannah areas flies may attack the collector at the rate of 2000 per hour or more, while at the other extreme much lower biting rates of the order of 10-20 per man hour may be the normal situation in forest areas.

Despite the recognized limitations of having to rely on a single sampling or study technique for adult *S. damnosum*, the man-biting rate remains the main study method for establishing behaviour patterns and determining flight range. For the last few years it has also been the sole method available to the WHO for assessing the impact of the larvicide-based control programme on the adult *Simulium* population (Walsh and co-workers, 1978; WHO, 1977).

Acute awareness of this unsatisfactory state of affairs marked a particularly interesting study in 1968-69 in the Cameroons (Disney, 1975), a long established centre for scientific studies on the epidemiology of onchocerciasis (Lewis and Duke, 1966). Previous studies in that region had made use of the systematic collection of man-biting records in order to define the seasonal fluctuations in numbers of *S. damnosum*. A closer examination of the situation revealed that conditions in different localities differed so much from each other as to make it unrealistic to define seasonal abundance of *S. damnosum* for the region as a whole. Not only did certain localities show distinctly different seasonal fluctuations, but also the fly density at any one place and at any one time was not necessarily related to the abundance, or even the detectable presence, of *S. damnosum* larvae and pupae in the local rivers. It appeared likely that in some of these villages the population of biting flies is probably derived from several breeding sites, and that at different times of the year different sites could be of greater or less importance as a source of adult flies.

In order to examine this possibility eleven different sites were selected in the middle region of the drainage basin of a particular river. Each month a uniform type of 'collection zone', with three palm fronds as artificial substrate, was intensively searched and collected for *Simulium* pupae — these pupae being readily identifiable to species — and the numbers tabulated as in Figure 7.6. These results showed that *S. damnosum* breeds in all the rivers sampled although it may not use a particular site for the greater part of the year. It was clear that each site had its own particular phenology curve (e.g. between sites IV and V) and that there was no consistent relationship between these curves and the regional rainfall pattern. When the seasonal abundance curve for adult *Simulium damnosum*, obtained from man-biting data, was compared to that of pupae in the adjacent parts of the river (Site V), there was found to be no regular correspondence (Figure 7.7). While the peak of pupae in September and October corresponded with the peak of biting flies, there is no peak of pupae to correspond with or pre-date the surge of biting flies in March and April. The interpretation of this is that the high proportion of flies biting man in this earlier part of the rainy season is *not* derived from the nearby river but from other rivers or localities where such increases in pupal density were recorded in the month of February, e.g. stations II, III and IV (Figure 7.6).

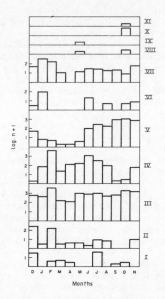

Fig. 7.6 The number of *Simulium damnosum* pupae collected in the survey. I-XI are the collecting sites. (After Disney, 1975).

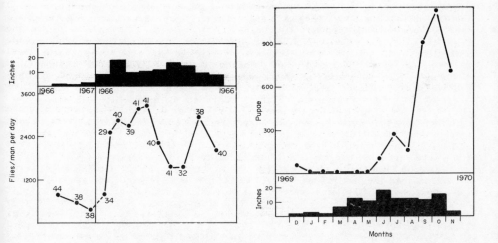

Fig. 7.7 Comparison of phenology curves of *S. damnosum* at Bolo derived from numbers of flies caught biting man during 1966 and 1967 (left) and the numbers of pupae coillected in the river at Bolo during 1969 and 1970 (right). The rainfall is also plotted. The numbers on the curve for biting flies indicate the percentages of flies parous. (After Disney, 1975).

A further implication from this is that not only is there a seasonal change in the breeding sites of the flies but also a seasonal change in the mean flight range of flies invading the village.

Collection of *Simulium damnosum* attracted to human bait has also been extensively used in marking-release-recapture experiments with this species. The points raised in the Cameroons work regarding the origin of *Simulium* population invasions according to the varying productivity of different rivers is only one facet of an increasingly important problem. There are many records from previous years indicating that *S. damnosum* may fly long distances, and this feature is now vital to the Onchocerciasis Control Programme (OCP). It is becoming imperative not only to define more precisely the conditions under which these invading swarms occur, both with regard to massive output of adults from particularly favourable breeding sites — possibly in inaccessible stretches of river or rapids — and also with regard to the various meteorological conditions which favour long-distance flight of the adults. In addition, knowledge of the composition of the female swarms with regard to physiological age (parity) and rate of infection with filarial parasites is an essential requirement for evaluating the practical significance of such invasions and in correctly forecasting their outcome. Two recent examples of marking experiments in West Africa illustrate the scope of such work.

The first experiment was carried out in the White Volta River Valley in the northern and upper reaches of Ghana (Thompson and co-workers, 1972). This is a northern Guinea savannah area with marked seasonal climatic variation, rainfall normally being confined to the period from April to October. In order to build up a high population of adult flies for marking and release, numbers of wooden sticks were suspended across breeding channels in the river and maintained there until they had accumulated large larval and pupal populations. These were then placed under a large emergence cage constructed over the river, and emerged flies caught by a man inside. Both males and females were collected and marked, but only females were included in the counts.

The dye marker was a fine, highly-coloured powder (Sudan III), mixed with flour and applied to the external surface of the fly. Apparatus was designed to apply the powder by blowing the flies around in a turbulent cloud of powder so that flies were exposed from all angles and powder was applied with sufficient force to penetrate deeply into the bristles.

It was found convenient to treat 100 flies at a time in this way when flies were emerging in large numbers, and 50 at a time otherwise. Twelve catching stations were chosen for accessibility and for their yield of high catches of *Simulium damnosum* in the past. At these points, all captures of adult flies were made by the normal human bait collection method. Recaptured flies were tested with acetone to detect the dye.

Over a two-month period from March to May nearly 60,000 flies were collected from the emergency cage, marked with dye and released. From 1st April over 22,000 flies were caught over the following three months but only a single marked fly was recovered, this recovery taking place at the nearest collecting site to the marking point, a distance of 17 km direct and 27 km by river.

This experiment highlights the many difficulties in this type of investigation under savannah conditions. While yielding so little of a conclusive nature, the experience did, however, suggest improvements of technique of possible value in later work.

In the Cameroons another mark-release-recapture experiment was carried out in the forest zone in order to determine the reduction in *S. damnosum* biting density with increasing riverine distance from the marking point (Thompson, 1976a), and to study the dispersal of flies along a main river as compared with a small tributary and with a road running at right angles to the river.

In order to obtain adequate numbers of adult flies for marking, a technique different from that used in Ghana was employed. Several men seated near the marking site caught unfed flies as they settled on their legs. At 15-minute intervals all flies caught were pooled for marking by fluorescent 'Day-glo' pigment, the colours used being red, orange and yellow. In the recapture programme each station was manned by 1–5 men who caught alighting flies over periods of several hours each day. Collections were examined 200 at a time by UV light in order to detect marked flies.

In addition to this, a rather different experiment was tried out in which the female *Simulium damnosum* attracted to human bait in the first stage of the test were allowed to engorge on their human hosts, all of whom were carriers of the infective stages of the *Onchocerca* filariae. While these flies were engorging, a tiny spot of paint was applied to the upper surface of the thorax. When the blood meal was complete they were allowed to fly away. Different colours indicated different dates of release. In three separate releases over 5,000 engorged females were effectively marked and released.

With regard to the marking experiment with unfed females, the number marked and released in seven days extending over a period of one month was 30,000. In the experiment on the large river 330 marked flies were recaptured, while 54 were recovered in the small river experiment. When recaptures were plotted according to distance from the release point, a dispersal pattern of flies over a period of six days could be estimated as follows: for every 1000 flies of this population found biting at the starting point during this period, 200 will be found during the same period at a distance of 34.1 km (100 upstream, 100 downstream), 20 will be found at a distance of 68.4 km (10 upstream, 10 downstream), and — by extrapolation — 2 will be found at a distance of 102.7 km (1 upstream, 1 downstream).

Observations on the small river indicated a much more rapid decline in dispersal than on the large river, and this decline was even more marked on the road running away from the river.

In the experiment with marking engorged females, two of these were recaptured at a collecting point 24 km away four days after their release in the gorged condition. These flies must have covered this distance either as gravid flies in search of an ovipositing site and/or as parous flies hunting for the next blood meal. The indications are that parous flies are no less capable of dispersing along large rivers as are the nullipars. A parous fly recaptured at this 24 km collecting station on the 7th day was presumably returning for its second blood meal after that taken at the time of marking (Disney, 1968, 1970). As it contained infective larvae of *Onchocerca*, this also demonstrates that it is possible for an infected fly to travel this distance.

Mark-release-recapture experiments, even under the most favourable conditions, have a discouraging tendency to reveal at best only a tiny fragment of the whole complex canvas. They do provide proof that, under certain conditions, flies can travel a considerable distance. What they cannot answer is whether these long flights are comparatively rare or whether they are really indicative of massive migration or dispersal sufficient to invade and recolonize remote rivers and river basins. Perhaps the most convincing evidence of mass flight over long distances is when high populations of *Simulium* suddenly appear in areas quite incompatible with the productivity of local larval habitats or, better still, at periods of the year when seasonal breeding streams and rivers in the invaded area have dried up. Studies on situations of this type in many different parts of the world have all contributed over the years to the long-established conclusion that blackflies can fly, or be transported on favourable winds, for long distances.

The rapid extension of larvicidal treatment of *Simulium* larval habitats in the OCP, since its inception in the Volta River Basin in 1974, has provided unique opportunities for revealing the extent of invasion potential (Davies and co-workers, 1978;   WHO, 1977, 1978).   As the progressive control operations have extended outwards they have created ever-widening areas where larval control has reduced the local man-biting population of adult flies to negligible proportions.   Nevertheless, massive invasions of *S. damnosum* have been regularly reported which cannot be explained by any local failure in control or by the existence of undetected local breeding foci.   The main study centre for this phenomenon has been in two of the great river systems of the Ivory Coast, the Bandama and the Leraba (Figure 7.8), marking the south-west extension of the control project centred in the Volta River basin.

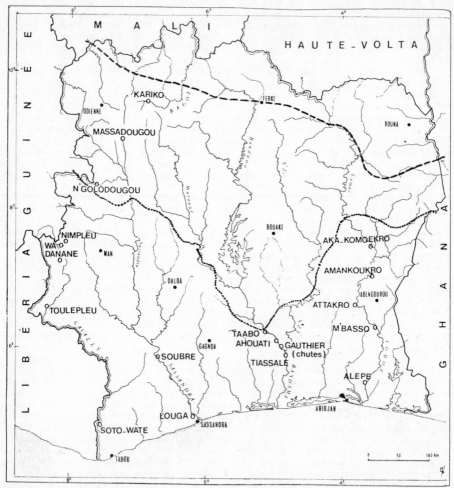

Fig. 7.8 Distribution of study areas in the Ivory Coast for the *Simulium damnosum* complex.   Upper dotted line shows boundary between Guinea savannah and Sudan savannah.   Lower dotted line shows border between forest and Guinea savannah.   (After Quillevere and co-workers, 1977).

After three years study it was clear that re-invasion of the controlled area was liable to occur every year, with increased biting rates throughout the months of June, July and August, coinciding with the rainy season. At first it was thought that forest areas outside the project limit were the origin, but larvicidal treatment of suspect forest foci failed to reduce invading populations. This was confirmed when improved techniques for the identification of the *Simulium damnosum* complex had progressed to the extent of being able to identify the invading flies as savannah cytotypes, not of forest origin. The south-west to north-east direction of the invasions corresponded to the direction of the prevailing winds and strongly suggested that invasion was wind-assisted. Large numbers of females were found to travel 200-250 km, and smaller numbers up to 300-400 km, from source. The invading female population is characterized by the high parous rate, the large majority arriving after a blood meal and ovipositing within the con-trolled area before taking a further blood meal. It was concluded that any area situated 250 km downwind from major breeding sites is open to re-invasion, possibly by infected flies. While the main capture or sampling method throughout these flight-pattern studies continued to be a network of man-baited capture stations, additional methods were introduced in the form of aluminium sheets or 'plaques', especially attractive to gravid females, and suction traps for collecting adults in flight. The use of these will be discussed below.

From all these reports it appears that, despite the wide range of techniques developed by *Simulium* workers in general, studies on adult flight patterns of *S. damnosum* are still largely dependent on the interpretation of standard collections on human bait, and the recording of densities on a man/hour basis (Quillevere and co-workers, 1977, 1978; Walsh and co-workers, 1978). Until comparatively recently other methods of capture or trapping adults had only made a negligible contribution to studies on behaviour and ecology of this species (Le Berre, 1966). Possible reasons for continued reliance on the man-biting rate were its simplicity of operation, the fact that it lends itself to standardization of catch and, perhaps above all, its high yield in so many areas of West Africa. In the vicinity of productive breeding grounds — particularly in the savannah zone — catches up to 2,000 flies per man hour are not unusual, and it is likely that even greater densities, beyond the maximum capacity of the bait/collector, have occurred. At the other extreme, in many forest areas much lower biting rates of the order of 10-20 per man hour are the normal experience, even in localities where a very high output of adults from known local breeding foci is known to occur. In areas under effective control the biting rate may fall to 1 or 2, or to zero, and it is at this low end of the scale that the question of the efficiency or sensitivity of this sole technique as a measure of adult popul-ation density becomes debatable.

Undoubtedly, the greatest gap in knowledge about the normal flight activities of *S. damnosum* has been due to the almost complete failure to determine the normal outdoor resting places used by the females between the period of engorgement and oviposition. This failure has not been due to lack of effort on the part of generations of experienced entomologists, but is a measure of the elusive nature of these small insects combined with the sheer physical difficulties of searching more than a fraction of the available resting sites in heavy forest or vegetation cover. In one locality in the Sudan savannah region of the Cameroons a systematic search of riverside vegetation yielded over 1,000 resting blackflies of several species, but of these only a small proportion had gorged, the total yield of *S. damnosum* itself being three (Disney and Boreham, 1969).

In another intensive search of vegetation at a dam site in north Ghana, where one would expect very high fly densities, results were more encour-aging (Marr, 1971). Daytime searches of vegetation, mainly carried out from

16.00 hours until dusk (18.15 hours), yielded over 4,000 adult *S. damnosum*, the great bulk of which were recently emerged males and females. However, from 17.30 hours onwards with approaching dusk larger numbers of gravid females were taken at the same sites, this concentration coinciding with peak ovipositing activity. The yield of engorged females was again discouragingly low, with only ten female *S. damnosum* in 16 days work. Extension of these systematic searches of vegetation to two other locations of high *S. damnosum* production on the Red Volta and the Sissili River only yielded a few females, including a single blood-engorged. In those two sites the density of the vegetation probably contributed to the arduous and unrewarding efforts.

The studies at the dam site did, however, provide a little insight into likely flight patterns of newly-emerged *S. damnosum* females. Bordering the spillway and drainage channel *Simulium* were found resting mainly on the long grass; but away from the water's edge (10-20 m) the flies were mainly 7-10 ft up on guinea corn, millet or bushes. On a tree 37 m from the spillway flies were found only above 16 ft (5 m) from the ground, where they could be seen flying just above bushes and trees before landing on the leaves. It seems likely that it is these upward flight movements which expose the flies to aerial currents and winds, enabling them to disperse over long distances.

In this connection the observations of Canadian workers in the course of work previously described (page 140) is very pertinent (Wolfe and Peterson, 1960). In those studies, mainly on *Simulium venustum*, flies were actually observed moving to night-time resting places in the tree tops, and were again observed at dawn flying down from the tops of the forest cover. Following this up, observers climbed trees at night, shook the vegetation and collected the disturbed blackflies by sweeping around with nets 20-30 ft above ground.

The problem was further investigated in the Vea dam site in north Ghana, using trapping techniques rather than direct collection by aspirator (Walsh, 1972). Sticky traps were constructed of plywood 1 ft square (0.09 m^2), painted dark green and coated with a sticky substance (Beacon Bird Repellent). The boards were fixed by wires to the trunks and main branches of five trees at differing distances from the spillway. Traps were set at different levels from 0.3 ft above ground to a maximum of 9.2 m. Over a period of about six weeks traps were examined daily, but the yield was disappointingly low; the average catch was one fly per five trap nights, and the bulk of the captures were newly emerged. Apart from the significant finding that *S. damnosum* may occur in the canopies of the tallest available trees, the virtual absence of blood-engorged females from all samples failed to provide any real clue to the nature of the normal resting places, by day or by night.

While there is no evidence to suggest that freshly-engorged females tend to congregate together in some as yet undisclosed type of resting site, the characteristic aggregation shown by *S. damnosum* females prior to oviposition (Muirhead-Thomson, 1956) is a feature which might well be used to develop a sampling method supplementary to the standard man-biting captures. Reference has already been made to the peak arrival of gravids in waterside vegetation at the Vea dam in Ghana, and it appears that this concentration of flies and flight paths plays an important part in two more recently developed sampling techniques.

In the first of these, developed by French workers in the Ivory Coast (Bellec, 1976), a very simple type of trap was designed consisting of a plain sheet of aluminium, 1 metre square, which is treated with adhesive and placed flat on the ground in the immediate vicinity of the normal bed of the river where there are rapids. Two to six sheets or 'plaques' are placed at different points along the watercourse on the same day. On 57 trap-days in the dry season over 21,000 *S. damnosum* were trapped in this way. The number of adults per day per sheet showed wide variations from 16 to over

2,000. Compared with man-biting rates in the vicinity, the sheet traps sampled a higher proportion of older females. About 11% of the total females captures were in the gravid state. The possibilities of this new technique are being vigorously followed up, and no doubt new information will soon outdate this brief progress report.

The second method arises from renewed interest in the possibilities of light traps for *S. damnosum*, despite discouraging earlier trials. As all available information indicates that *S. damnosum*, like most of the Simuliidae, is predominantly a day biter and only active by day, the light trap would seem *a priori* unlikely to succeed. However, intensive studies on light trapping in Scotland from 1960 onwards proved surprisingly productive (Davies and Williams, 1962), indicating considerable night-time activity of all physiological stages of adult females, including freshly engorged, and also yielded adults of species not captured in any other way. Despite these encouraging results, the first trials with light traps in West Africa gave disappointing results, and interest remained low until the whole problem was vigorously re-examined in north Ghana from 1970 onwards.

Trials using both ultraviolet and fluorescent light were carried out in 1972-75 at active breeding sites at Sugu in the White Volta, at the Vea dam near Bolgatanga, and on the Red Volta and Sissili River (Walsh, 1978). Some of these traps operated throughout the entire night while others were restricted to the first five hours of darkness. The tests proves to be very successful and particularly productive in the rainy season. Very large catches of *S. damnosum* were taken in the first 2-3 hours of darkness, after which the catch fell off rapidly. Examination of over 5,000 *S. damnosum* caught in the early hours of the night showed that 99.5% were females, of which at least 57% were gravid, and possibly more because of the unknown number of flies ovipositing in the collecting bag and subsequently being classified as 'parous'. Later in the night, the proportion of males and nulliparous females increased.

These light trapping results provided strong evidence of nocturnal activity, in both the dry and wet seasons. The particular member of the '*S. damnosum*' complex involved was identified as *S. sirbanum*. Compared with the conventional human bait sampling method, light trapping proved relatively less efficient with a mean catch of 15 *S. damnosum* per night, as compared to about 250 flies per man-hour taken by a vector collector in the locality.

It appears that light is particularly attractive not only to gravid females in the early hours of darkness, but also to newly-emerged flies of both sexes later in the night, evidently shortly after emergence. Unfortunately there was little evidence that this attraction extended to engorged females, only six of which category were recovered in the entire operation.

Intensive light trap studies were again taken up in Ghana from 1978 onwards, in which the well-tried design of Monks Wood trap was modified in such a way that the fluorescent light tube could be operated either continuously or repeatedly flashed on and off at different rates (Service, 1979). Two light traps were set up about 7 m apart and operated from 19.30 to 06.00 hours. One of the traps had an ultraviolet tube set for a one-second flash rate, while the second trap used a continuous white light. Later, one of the traps was used with a continuous ultraviolet light.

A total of nearly 15,000 *S. damnosum* complex females were taken on four nights, the highest collection for any one night being 6520. The relatively high proportion of gravids (12.2%) and the fact that many of the unfed females examined by dissection proved to be parous, point to the conclusion that these traps attracted blackflies prior to or soon after ovipositing. They were not catching newly-emerged females, and only negligible numbers of males were taken.

By the time this investigation was carried out progress in the definition of the *S. damnosum* complex had reached a stage where it was possible to

identify these particular members as *S. squamosum*. The successful results
of light trapping in this particular locality, the Pawnpawn river, taken in
conjunction with the comparative failure of the same techniques in preliminary
tests on the Volta river about 40 km away, suggests that trap location is a
very important factor and that a great deal more research will be required
in order to fully exploit this promising technique.

The unsatisfactory but unavoidable dependence at the moment on man-
biting data as a measure of *Simulium* density has prompted critical studies
in Cameroun on the factors which attract *Simulium damnosum* (Thompson,
1976b,c,d). It was hoped that such studies might pin-point certain vital
factors which could be put to practical use for the design of a mechanical
trap to replace live human bait as attractant.

Experiments were designed to evaluate the relative importance of visual
stimuli (sight), olfactory stimuli contained in human breath (exhaled breath),
and olfactory stimuli emanating from the skin surface (smell). The problem
was approached in two ways: firstly, by presenting as bait a man emitting
all the attractant factors except the one under consideration; and, secondly,
by attempting to isolate as far as possible, and present as bait, the factor
under consideration. In this method of experimental approach this investigation shares many of the features of concurrent studies on tsetse in Rhodesia
(page 98) but it is clear that the methods were arrived at quite independently.

Traps were specially designed for this purpose. The first of these was
a triangular frame 120 cm high, and 44 cm each side. Two sides of the
column were covered with fine mesh, while the third was provided with a
close series of inclined slats. The trap was suspended from the branch of a
tree with the open bottom 10 cm above the ground. The operation of the
trap was based on the belief that flies would travel upwind towards the bait
inside the trap, and finally land on the slats. Moving upwards they would
reach the top of the slats and then find themselves inside the column where
they would be attracted to the two net-covered sides through which the daylight penetrated. From there they would continue to move upwards towards
the top of the column. This column was topped by a netting cage constructed
in such a way as to leave a 10 mm gap between the outside wall and an
inverted net tetrahedron suspended in position with twine (Figure 7.9). This
downward pointing tetrahedron was kept in a taut position by means of a
small weight, ensuring that the narrow 10 mm gap was kept rigidly open to
allow ascending flies to move upwards into the cage, where they were trapped
and could be removed through side sleeves.

The second trap (Figure 7.10), large enough to contain a man, was
called the enclosure trap, and was based on the slat principle with the slats
adjusted so that the man was not visible from outside. This trap was used
to observe the number of flies approaching a man emitting smell and exhaled
breath stimuli, either separately or combined, in the absence of visual
stimuli.

In the first series of experiments with the slat trap, comparisons were
made between (i) the empty trap, (ii) a trap provided about 20 cm in front
of the slats with a rubber hose outlet through which a man, seated 30 m
away in thick bush, could exhale; (iii) a trap provided in front of the
slats with recently-worn trousers; and (iv) a trap provided with both exhaling hose and trousers. Over the same period two collectors 100 m downstream
made conventional man-bait collections. The results are shown in Table 7.1.
From these figures it can be seen that exhaled breath and worn trousers on
their own attracted no more flies than the empty trap, but that the catch was
increased roughly seven-fold when exhaled breath and trousers were presented
together. But even with this combination, the catch was less than 10% of
the comparison man-bait catch in the open.

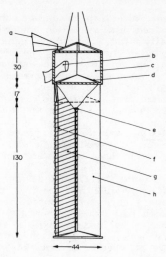

Fig. 7.9   The slat trap.   (a) "Rudder";
(b)   sleeve;   (c)   fine-mesh ("sand-fly")
netting; (d) 10-mm entry gap; (e) weight;
(f) slat holder;   (g) slats (43 x 5 cm);
(h) 1-mm mesh netting.   Measurements are
in cm.   See   text for   further   details.
(After Thompson, 1976).

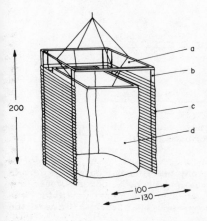

Fig. 7.10   The enclosure trap, with one
side removed.   (a) 1-mm mesh netting;
(b) polythene (transparent);   (c) slats
(130 x 5 cm).   Measurements are in cm.
See text for further details.
(After Thompson, 1976).

Table 7.1 The numbers of *S. damnosum* captured over 12 hours in slat traps
with various baits on successive days. (After Thompson, 1976).

	No bait	Exhaled breath	Worn trousers	Exhaled breath and worn trousers
Trap catch	54	42	30	256
Vector collector catch	3365	2728	3224	2652
Trap catch as proportion (%) of vector collector catch	1.6	1.5	0.9	9.7

In a second series of tests, using the enclosure trap, a human subject not visible from outside sat inside and compared his attraction under two different conditions: firstly, exhaling normally and, secondly, exhaling through a rubber tube with its outlet 30 m away in thick bush. The results showed that when visual stimuli are reduced, human smell alone attracted four times as many flies as the unbaited trap, but when exhaled breath and smell were combined the catch increased by a further factor of four.

In a third series, the attraction of human exhaled breath was tested when visibility of the human bait was optimal, in contrast to the test above. This was done by one of the exposed bait men exhaling through a rubber tube with its outlet 50 m away. The results showed that the removal of exhaled breath did not remove the attraction of fully-visible humans, and these results applied to both environments in which these tests were carried out, i.e. forest and savannah.

A further experiment, using a fan trap (Figure 7.11), was designed to study the attraction of a human bait when fully visible and moving round from time to time, as compared with one stationary and partly concealed. In each case a further comparison was made between bait exhaling normally and bait exhaling through a rubber tube or hose to a distant exit. These experiments showed that the hidden stationary man always attracted fewer flies than the moving exposed man, irrespective of whether or not they were exhaling normally.

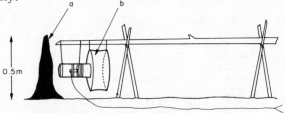

Fig. 7.11 The fan trap.   (a) Bait cloth;   (b) collecting cage. (After Thompson, 1976).

In further experiments using a transparent plastic enclosure, the human host was easily visible, but the exhaled breath and body odours were confined to the air-tight enclosure. At a time when biting density outside was of the order of 1000–2000 flies per hour, the visible but odourless bait in the plastic enclosure was completely non-attractive.

By means of a fan trap, various choices of washed versus unwashed garments were presented as stimuli, and these showed that worn clothes attracted over ten times as many flies as unworn clothes. In both cases the attraction was greatly enhanced by the addition of CO_2, up to 18 times greater in the case of the worn clothes. The findings with regard to the strong attraction of human body odours also provided an explanation of the observation that fan traps which had been used for several consecutive days showed higher and higher catches in the controls. By wearing gloves and protective clothing, and washing and sterilizing the equipment, it was shown conclusively that high control catches stemmed from human handling contamination.

These experiments also revealed differences between the reactions of flies in the Sudan-savannah zone of the Cameroun, and the forest zone. In the savannah, the removal of exhaled breath and the masking of human odour failed to produce any substantial reduction in the number of flies attracted to human subjects. This suggests that in that zone, some other stimulus, probably sight, is a major factor in attraction. In this case it seems very likely that the differences in behaviour reaction in the two zones could be attributed to the likelihood that two different cytotaxonomic species of the *Simulium damnosum* complex were involved, the savannah flies being *S. damnosum* and those in the forest *S. squamosum*.

These and other results suggest that different members of the *Simulium damnosum* complex may, in their host-locating activities, respond to different kinds of stimuli emitted by the human host. The major role of human body odours as attractants in the forest zone experiments was further analysed by comparing the attractiveness of clothing worn on different parts of the body, using the fan trap technique. In these tests trousers were found to be 5.5 times more attractive than shirts. This led to an examination of the attraction of various body fluids and exudates, of which sweat emerged as the major attractant, particular sweat from below the human waist line. Attempts to isolate and identify these particularly attractive chemical compounds in human sweat failed to pinpoint the active ingredient, possible because such chemicals did not react with the range of organic solvents used in extraction. Whatever this active compound is, flies seem to be able to detect extremely small quantities as evinced by the experiments with fan traps 'contaminated' by human handling. Should these studies lead eventually to the extraction of this powerful compound (or compounds), then there would be a real possibility of using that substance to design a standard trap which would obviate the continuing need for human bait and man-biting indices as the sole measure of *Simulium* activity and abundance.

One of the most striking features emerging from the extensive work on *Simulium damnosum* is the paucity of information about blood feeding activities and preferences apart from man. It has long been recognized that other hosts must exist, especially in some forest areas where comparatively low man-biting rates are incommensurate with the high adult populations indicated by high pupal densities in local breeding streams (Duke and co-workers, 1966; Garms, 1973; Lewis and Duke, 1966). The role of the larger domestic animals has not until very recently been critically assessed by direct capture methods, and in this connection the study methods used with the *Simulium* vectors of ochocerciasis in central and south American countries are particularly illuminating. Of the three main species involved in the Americas, one, *Simulium ochraceum*, is considered to be mainly a man-biting species, while the other two, *S. callidum* and *S. exiguum*, bite man and domestic animals more indiscriminately. The disclosure in recent years that in those areas *Simulium* is also involved in the transmission of viral disease, namely Western Equine Encephalitis (WEE) and vesicular stomatitis, has renewed interest in their behaviour and ecology.

This is well illustrated by recent work in Colombia, where host preference studies have been carried out not only in the higher altitude coffee-growing districts where onchocerciasis is prevalent, but also in lower localities associated with outbreaks of WEE, and in rain forest areas (Guttman, 1972). The capture method adopted was the daytime collection by aspirator of biting flies attacking cow, horse and four locally-hired fly boys. The concentration of biting flies on the undersurface of the animals made it easier for one collector to cope with heavy attack rates. Two of the human bait boys were stationed near the animals in an open corral, while the other two were stationed in dense vegetation at least 30 m away. Collections were made for 20–30 minute periods throughout the day.

Parallel trials on the attraction of small caged mammals and birds by methods which had been successful in North America proved unproductive and were abandoned. Biting density on man ranged from 0.57 per man hour (dense vegetation) to 19.2 (in the presence of horse and cow in the open). On cattle and horses the numbers ranged from approximately 70 to 151 per hour. The species in those tests were predominantly *S. exiguum* (66% of total) and *S. callidum* (32%).

The records confirmed that in the three environments sampled, *S. exiguum* and *S. metallicum* were the most common man-biting species, but when fly boys, cows and horses were stationed together, the majority of the biting population fed on horses and cows with only a few on the human host. Clearly, these two species would appear to be less markedly anthropohilic than '*S. damnosum*' in Africa. Nevertheless, it would be useful to have more precise figures about the reactions of *S. damnosum* to such an experimental choice of mammal hosts, expecially in areas dominated by different members of the complex.

It has long been recognized that birds are the most likely alternative hosts for *Simulium damnosum*. This is not only based on odd observations on chickens but also on the premise that many of the filarial infections found in *S. damnosum* appear more likely to be of avian origin than human origin (Garms and Voelker, 1969). One of the few systematic attempts to study this experimentally, especially without the interference of human presence, has been carried out in the Cameroons (Disney, 1972). The basic principle adopted was essentially the same as that used successfully by North American workers, namely lowering a cage periodically over freely-exposed bird bait.

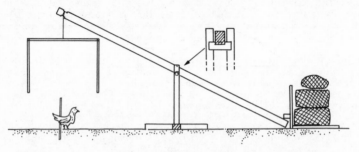

Fig. 7.12 Apparatus for catching flies attracted to a chicken. Suspended cage to left; pivot in centre — with enlargement (as viewed from position of chicken) above; anchor weight to right. (After Disney, 1972).

The method used in the Cameroons is illuatrated in Figure 7.12, and is self-explanatory. A significant feature is that the bait bird (chicken) is tied to a stake and is not caged. The cage was repeatedly raised and lowered in the course of an hour, and catches were continued throughout the day. Out of over 1,400 flies caught in this way, 526 were *S. damnosum*, a sufficient number to be broken down into hourly captures and used to plot the diel biting cycle (Figure 7.13). In this way it could be shown that the daily biting pattern of *S. damnosum* on its avian host closely follows that already established on the human host in that region. Morning biting rates increased steadily to reach a plateau extending from afternoon to dusk. This is in striking contrast to the biting pattern of another common species, *S. uni-cornutum*, which is never taken on human bait, and which has a high early morning peak of biting on chicken bait and a smaller peak before dusk.

All captured flies were classified as to parous and nulliparous. Of particular interest is the parous biting rate on chicken host which follows a pattern significantly different from that on man (Figures 7.14, 7.15). The interpretation put on this is that the minor peak of flies around 09.00 hours represents those which have failed to obtain a blood meal the previous day, and which are biting earlier in the day than the corresponding wave of nullipars. The main peak, 11.00-14.00 hours, starts later than the corres-ponding wave of parous flies biting man because when they begin to seek

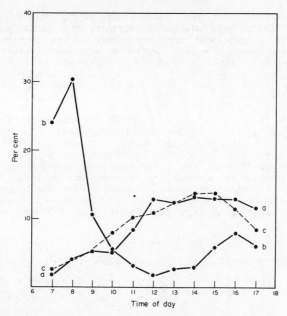

Fig. 7.13 Biting cycles of blackflies caught on a chicken at Bolo compared with biting-cycle for man-biting S. damnosum. (a) *S. damnosum* on chicken. (b) *S. unicornutum*. (c) *S. damnosum* on man. (After Disney, 1972).

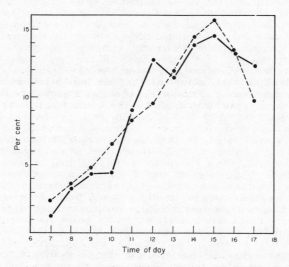

Fig. 7.14 Biting-cycles of nulliparous *S. damnosum* at Bolo. Solid line = flies caught on chicken; broken line = flies caught on man. (After Disney, 1972).

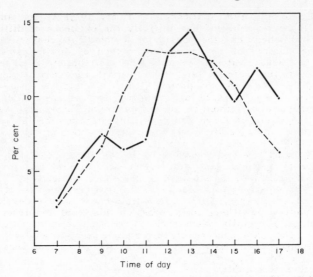

Fig. 7.15   Biting-cycles of parous *S. damnosum* at Bolo.
Solid line = flies caught on chicken;   broken line = flies caught on man.
(After Disney, 1972).

their blood meals the parous flies are very anthropophilic.   Lastly, as the
factors leading to curtailment of host-seeking activity come into operation,
the flies become more catholic in their choice of hosts so that a minor wave
of chicken biting flies is recorded around 16.00 hours after the numbers have
started to decline sharply.   Interpretation of these observations has to take
into account the already established fact that the biting pattern of nulliparous *S. damnosum* differs according to whether the environment is forest or
savannah (Duke, 1975a; Lewis and Duke, 1966).   In the former, biting by
nullipars occurs mainly in the afternoon, while in the latter it occurs mainly
in the early morning and late afternoon.   In both environments parous flies
tend to bite man mainly in the middle of the day.   Progress subsequent to
this work in clarifying the distribution of the newly-defined members of the
S. damnosum complex may, of course, reveal that the situation is further
complicated by more than one species of '*S. damnosum*' in the Cameroons
experimental area.

Somewhat similar study methods with regard to bird-feeding *Simulium*
have been used in another part of Africa, the Sudan, regarding a different
species, *S. griseicolle*, long known as a serious pest of domestic and wild
birds, and a biting nuisance on man (Bashir and co-workers, 1976).   The
bird bait capture method, which was one of three sampling techniques used
(see below) was again somewhat similar to the North American designs and
consisted of a box, with a roof window, which was lowered periodically over
the exposed host.   The inner surface of the glass roof-window was coated
with tree banding on which flies attempting to leave the box were trapped.

Defining the behaviour patterns of the *S. damnosum* complex.   Very
rapid progress in studying the different members of the *Simulium damnosum*
complex in West Africa has been made possible in the last 2-3 years   by
defining certain morphological features which enable adult females of at
least the six main West African species to be identified (Davies and co-

workers, 1978; Garms, 1978; Garms and Vajime, 1975; Vajime and Quillevere, 1978). It will be remembered that initially the existence of different members was firmly established on the basis of cytological characteristics of the larvae, in the absence of any consistent method of distinguishing the adult. Examination of larval material enabled the distribution of different members of the complex to be mapped, and their larval requirements clarified. But precise definition of behaviour patterns of the adult fly remained difficult. In most localities two or three different members of the complex were found to occur together, and although one or other of these might appear to be dominant on the basis of larval surveys, investigators could not assume that adult captures in that locality were necessarily those of the dominant larval population.

Intensive work by the French team in the Ivory Coast has now revealed reliable means of distinguishing adult females, and with this great advantage it appears that from now on studies on behaviour patterns of 'S. damnosum' can be directed separately to each of the member species of the complex (Quillevere and co-workers, 1977, 1978). In such a fast developing subject it would be unduly ambitious in a book of this kind to present information right up to date, especially when such information is subject to revision from time to time. The main concern of this review is to assess what progress in study methods has been made in keeping with the new knowledge of the S. damnosum complex.

Most studies on adults still continue to be based essentially on man-biting rates. These figures provide information firstly about the intensity of man-biting, secondly about diel biting cycles, and thirdly about the relative proportion of parous and nulliparous females at different periods of this biting cycle. The man-biting rates also enable a comparison to be made about the dispersal rates of each species from their main larval habitats, with particular reference to the proportion parous. In addition, increasing use has been made of animal baited traps or bait collections on chicken, sheep, rabbit and cattle as a measure of zoophily of different members of the complex. Finally, a record of infection rates with Onchocerca volvulus is used as a measure of the vector potential of each species.

The basic man-biting captures provided the bulk of the information about populations in the dry season and in the rainy season, and were also carried out in localities representative of different bioclimatic zones. Particular attention has been paid to comparisons of behaviour patterns relative to the savannah and the forest environment. These studies have so far revealed that the aggressive peaks of man-biting by the three different species of the complex included in the preliminary studies are largely determined by the daily variations in temperature. These peaks, mostly composed of nulliparous females, show that in general S. sanctipauli appears to be more sensitive to high temperatures than S. soubrense and S. yahense (Figures 7.16, 7.17).

These investigations, as well as those on zoophilism, still continue and it is perhaps too early to review solid progress. It seems fairly certain, however, that progress towards a better understanding of the behaviour patterns of the different member species will continue to be handicapped until methods are devised for sampling more representative sections of the adult population, and not just the biting fraction.

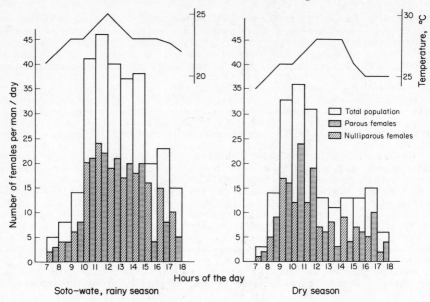

Fig. 7.16 Seasonal abundance (parous and nulliparous females) of *Simulium yahense* (*Simulium damnosum* complex) during the rainy season and the dry season in different localities in the Ivory Coast.   (After Quillevere and co-workers, 1977).

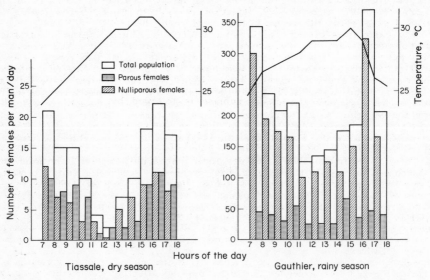

Fig. 7.17 Seasonal abundance (parous and nulliparous females) of *Simulium sanctipauli* (*Simulium damnosum* complex) during the rainy season and the dry season in different localities in the Ivory Coast.   (After Quillevere and co-workers, 1977).

iv.   Truck Trap Studies in England and the Sudan

One of the most interesting examples of techniques well established in one field of biting fly studies only receiving full attention in another field after a long interval is provided by the revived interest in truck traps for sampling Simuliidae.   Although the value of this 'non-attractant' capture method for insects in flight had been firmly established for nearly 30 years in connection with mosquitoes and biting midges, and had been the subject of trials on Canadian blackflies a few years after that time, it is only comparatively recently that its real possibilities with *Simulium* have been fully explored.   Additional interest is provided by the fact that these trials have been carried out successfully in two widely different environments and with different species, in Britain (Davies and Roberts, 1973) and in the Sudan (El Bashir and co-workers, 1976).

In England, the technique was redesigned specifically for the requirements of *Simulium* and incorporated a cone-shaped net mounted on top of a van driven along the road, and capable of filtering very large quantities of air — over 100,000 m^3 in half a day.   With the vehicle driven at a fixed speed of 30 mph (48 k/h) flying insects gathered into the net were funnelled into a pipe at the rear.   This pipe led into equipment which allowed the flow of insects to be divided into separate catches by changing the collecting tubes through which the air flowed.   The cone, 91.5 cm wide at its mouth, and the netting attachment, were designed in such a way as to prevent a build-up of air pressure in front, allowing the air to flow through without loss of flies.   An ingenious electrical system controllable by the driver ensured that collections made over each kilometer could be segregated in tubes on a rotating disc.

In preliminary trials with five runs totalling 220 km, 4681 *Simulium* were taken, mainly composed of three species.   Females in all stages of the gonotrophic cycle were obtained, including gorged flies, and it was also possible to define female distribution in relation to known breeding sites.

The truck trap experiments carried out independently by another group of workers in Sudan were of a simpler and more basic design in which a sweep net was held vertically through the slit in the roof of a saloon car cruising at 20 km/h.   There was no sophisticated funnelling or sorting mechanism, trapped insects simply being anaesthetized and stored.   The usefulness of this particular trial was enhanced by the fact that two other sampling methods were compared simultaneously, namely chicken-baited trap (see page 162) and suction trap.   The suction trap was electrically operated, with the air inlet 150 cm above the ground, the catch being automatically sorted into hourly collections.   Each time it was used the suction trap was operated at a constant selected speed for 24 hours.

Both these techniques revealed a biphasic biting pattern for *S. griseicolle* females, with the numbers starting to increase suddenly just before sunrise and reaching a relatively small peak between 07.00 and 08.00 hours. After a decline, the second and sharper peak occurred about 17.00 hours (Figures 7.18, 7.19).

A comparison of truck trap with suction trap showed that in the former the early morning catch recorded nearly equal numbers of both sexes, but that later in the day females greatly predominated (Figure 7.18).   The suction trap consistently showed a higher proportion of females throughout the entire day (Figure 7.19).   The sharp peak of flight activity recorded by both truck and suction trap towards the end of the day is duplicated by a peak of biting activity on chicken bait at the same period.   These and other observations all showed that the sharp peak of biting activity is also characterized by increasing non-selective attacks by this mainly ornithophilic species on various other available hosts including man.   Although this apparent change in host selection would appear to be a phenomenon reminiscent of that reported for other ornithophilic species in Canada, which turn

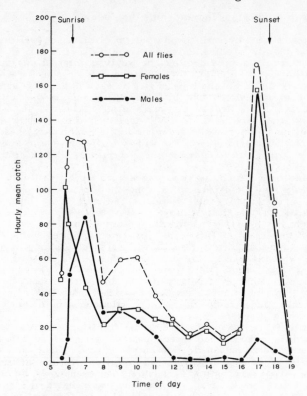

Fig. 7.18 Average catches of adult *S. griseicolle* plotted against time (vehicle trap). Average of five observations. (After El Bashir and co-workers, 1976).

to biting man with the onset of cold conditions in the autumn (Defoliart and Rao, 1965), the case with *S. griseicolle* is significantly different. Here it is a daily happening which may well be due to daily changes in climatic conditions, to an accumulation of teneral females through the day, or possibly to a local non-availability of the more normal avian hosts. Wherever suitable sections of motorable road exists, there seems every encouragement from this example for further trials of vehicle-mounted traps in other *Simulium* areas, including those in which species of the *Simulium damnosum* complex are prevalent.

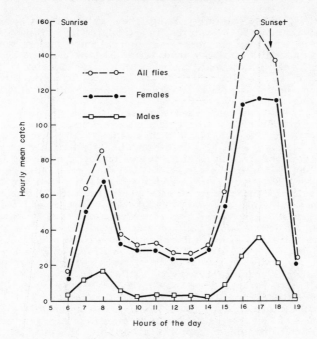

Fig. 7.19 Average catches of adult *S. griseicolle* plotted against time (suction trap). Average of five observations. (After El Bashir and co-workers, 1976).

v.   Australian *Simulium* Studies

In the last few years there has been a considerable resurgence of interest in the two main pest species of *Simulium* in Australia.   These are *Austrosimulium pestilens*, a troublesome pest of man, domestic animals and marsupials, and the closely related *A. bancrofti*, which is a man and live-stock pest in Western Australia but not in Queensland where the recent intensive research has taken place.

In Queensland, adults of *A. pestilens* are notorious for their ability to appear in huge swarms within a few days of the watercourses being flooded by heavy rains after many months in which these rivers have been bone dry. This phenomenon has led to disclosures about the remarkable ability of *A. pestilens*, laid in turbulent flood waters, to settle to the bottom of the river and survive the long periods of complete drought and desiccation of the river beds which follow the short rainy season (Colbo and Moorhouse, 1974). This work has a considerable bearing on similar, though less dramatic, situations in other regions of the world such as the savannah areas of Africa where normal breeding streams and rivers either stagnate or dry up completely in the long dry season.

However, from the point of view of the subject of the present book, the main interest in the Australian work must be the novel approach to capture and sampling methods for adults, used in establishing behaviour patterns of these two species (Hunter and Moorhouse, 1976).   Host-seeking females were captured in CO_2-baited tent traps (90 x 90 x 45 cm) tied at each corner to sticks or shrubs, these captures being then related to trap location and meteorological conditions. Captures in standard tent traps were compared with those in a tent trap provided with internal cloth baffles that reached to the ground. In addition, sticky traps were used in the form of clear plastic sheets, 30 cm wide, which were coated with 'Stikem Special'. These sheets were made rigid by tacking them to a box or to poles driven into the ground, catches being separately recorded from each 10 cm of height starting from ground level.

The CO_2 tent traps were used extensively to study the seasonal abundance of these two species in the river systems of southern Queensland, with special reference to the effects of floods and to the rise and fall of the rivers. Close studies were also made on the effect on fly activity of such meteorological factors as temperature, saturation deficit, insolation, degree of cloud, wind and barometric pressure. Catches were made with reference to the elevation of the capture sites, and their distance from the river; these revealed behavioural differences between these two closely-related species in that, while more *A. pestilens* were caught near the river, the numbers of *A. bancrofti* generally increased with elevation away from the river.

In order to define adult habitats more precisely, catches were recorded in four vegetation types using three capture methods: viz. CO_2-baited tent trap without baffles; CO_2 test trap with baffles; and sticky traps. The results indicate, firstly, the superiority of both baffled tent traps and sticky traps over standard tent traps without baffles and, secondly, the very low capture by all three trapping methods in a locality where a tall eucalyptus tree had shed its bark over the ground up to 4 m radius.

Direct visual detection of adult *Simulium* resting in tall grass led to an extensive series of net sweeps in a range of different vegetation types. At the same time a comparison of captures at different heights using CO_2 baited sticky sheets disclosed further behaviour differences in that *A. bancrofti* was caught mainly at heights of 20–70 cm, while *A. pestilens* was found at heights up to 3–4 m or more.

An extensive sweep-net collection among trees along the river's edge yielded very large numbers of *A. pestilens*, often in excess of 1000 per tree sweep. On some eucalyptus at the water's edge flies were found resting up to at least 10 m. These collections produced almost entirely females, including a range of physiological conditions such as freshly-engorged, half-gravid and gravid, and indicated that these are the sites where digestion of the blood meal and development of ovaries takes place. On this question of blood digestion, *A. pestilens* provided a unique contrast to other *Simulium* discussed so far in the extreme rapidity with which egg maturation takes place. While records for Canadian species indicate a period of 5–8 days, and for the African *S. damnosum* a period of 2–3 days, the whole process is completed in 24 hours with *A. pestilens*. This rapid ovarian development allows females to take a number of blood meals during their short lives, a fact which also enhances their potential as vectors of disease.

Of the many important new contributions to *Simulium* behaviour revealed by these trapping and sweeping studies, there is one of special significance arising from the work on host selection. *A. pestilens* voraciously bites both man and domestic animals including cattle, sheep, horses and dogs. *A. bancrofti*, on the other hand, does not attack man in Queensland, and is also attracted to animals to a much lesser extent. A comparison of the numbers of *A. bancrofti* attracted to animal baits, to animals plus CO_2, and to CO_2 alone, showed the superior attraction of CO_2 alone. This attraction was in no way enhanced by placing an animal in a trap already baited with CO_2. In the case of man the situation was very different, as the presence of man in a CO_2-baited trap had the effect of depressing the numbers of *A. bancrofti* attracted; the actual figure was 7.6 dropping to 0.6 flies per hour. Under similar conditions the attraction of *A. pestilens* to a CO_2-baited trap remained much the same, whether man was absent (20.8 flies per hour) or present (20.0 flies per hour). It appears that, as with the tsetse flies *Glossina morsitans* and *G. pallidipes* discussed earlier (page 88–89), man is not the natural host of *Austrosimulium bancrofti*, to such an extent that his presence has a repellent effect on this species.

CHAPTER 8

Biting Midges (Ceratopogonidae)

i.   Introduction

The biting midges (Ceratopogonidae) occupy rather an unusual and
anomalous position in any general assessment of vector biology and human
disease. While the biting habits of the females, and the preference for
mammalian blood shown by three important genera (*Culicoides*, *Lasiohelea* and
Leptoconops) has led to their incrimination as vectors of pathogens such as
filaria and viruses (Kettle, 1965, 1969c), none of these infections produced in
man are fatal or even of public health importance. The situation with
regard to domestic stock is rather different, and *Culicoides* are of considerable economic importance as vectors of Blue Tongue in sheep and African
horse sickness.

171

Despite their low priority as vectors of human disease, biting midges have been intensively studied in many countries, particularly in the New World, almost certainly due to the fact that in many regions they are major biting pests, seriously interfering with amenities and recreation in holiday and tourist resorts. In Florida, the Bahamas and many other parts of the Caribbean, sandflies are a major factor affecting the growth of tourism (Linley and Davies, 1971). Many of the secluded sub-tropical islands with attractive sandy beaches now being developed provide the very environment where persistent attacks by biting midges can seriously interfere with rest and relaxation, huge midge populations being produced from the extensive swamps and coastal brackish areas often associated with such places and in which the wet mud provides an ideal larval habitat.

Two of the most important species concerned in the New World are *Lasiohelea becquaerti* in the region from Florida south into the Caribbean, and *Culicoides furens*, a serious pest on the Atlantic and Gulf coasts of the United States. In attempting to measure population densities and fluctuations, and to establish some idea of the behaviour patterns of these biting pests, a range of conventional methods have been used by different research workers. In the United States, for example, there has been considerable emphasis on the use of light traps, and surprisingly little use of methods involving biting rates of landing rates on human bait. In contrast, the use of standardized human bait was the dominant technique adopted by the British workers in Jamaica in their surveys from 1965 onwards (Kettle, 1969a,b; Kettle and Linley, 1967a,b, 1969a,b).

ii.   **Biting Catch Studies in Jamaica**

In concentrating on biting catches and biting habits, the Jamaican team first considered the advantages and limitations of other current catching or sampling methods.   Light traps were dismissed because of their obvious ineffectiveness for diurnal biting species, and also because they do not attract all nocturnal species equally.   Suction traps, although producing apparently unbiased or non-attractant samples, depend on the insects flying at the same height as the mouth of the trap.   In addition, suction and light traps were considered relatively expensive and required sources of electricity as well as reliable operators.   For example, on the north coast of Jamaica biting catches in one locality indicated that two pest species, *Culicoides barbosai* and *C. furens*, were present in roughly equal numbers, but sticky traps in the vicinity produced only two females of *C. barbosai* compared with 506 *C. furens* females (Kettle, 1969a).   On the other hand, sticky traps *did* catch species not attracted to man, and therefore undetectable by human bait alone.

Ideally, bait observations would be carried out through the entire 24-hour period in order to include species active at different periods of that diel cycle;   but for routine practical procedures it was considered sufficient to select certain times at which particular species were known to be active. For example, both *C. barbosai* and *C. furens* were active at dawn, while *Leptoconops becquaerti*, poorly represented at sunrise and sunset, was best sampled 1½ hours after sunrise.

In selecting localities for biting catches, it was realized that differences in the environment (e.g. between shady and exposed sites) could be important, and that, in addition, open sites would be more adversely affected by strong winds than sheltered sites. However, open sites were selected not only because the 'sandfly' nuisance is associated with such locations, but also because in such places most elements of the microclimate would be fairly stable.

Five individuals were used in these human bait experiments, and it was observed that these individuals differed in their relative attractiveness to midges. Subsequently, therefore, adjustments had to be made to allow for that factor. It was also observed at an early stage that specific differences existed as to which part of the exposed body was a preferred biting site. For example, *C. furens* concentrated its attack on the back of the calves, while *C. barbosai* attacked the arm of bait in the same seated position on the ground. It was therefore necessary to introduce a regime whereby either two of the collectors exposed their legs while the other two exposed their arms, or each individual offered legs and arms alternately throughout each trial (Kettle and Linley, 1967a). These experiments showed that *L. becquaerti* had the strongest preference for a particular limb, being four times as numerous on the leg as on the arm, *C. furens* was about 1½ times as numerous on the leg, and *C. barbosai* only two-thirds.

Preference for the lower limb might be explained in part by the fact that the diurnal habits of *L. becquaerti* coincide with periods of high wind, when even the slight reduction in air movement between the level of the arm and that of the leg of a human sitting on the ground is sufficient to affect biting activity. However, at night, when such air movement is minimal, there are still trends to lower biting by these species. The predominantly diurnal biting of *L. becquaerti* is illustrated in Figure 8.1, which shows the two main peaks of biting between 08.00 and 09.00 hours in the morning, and again in the late afternoon before sunset (Kettle and Linley, 1967a).

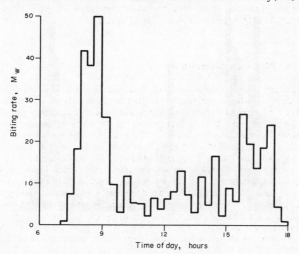

Fig. 8.1   The biting cycle of *Leptoconops bequaerti* on Florida beach, January 1960. (After Kettle and Linley, 1967b).

When these observations were extended through the entire 24-hour period, and weekly throughout a complete year in order to include all seasons, it was found that many variables had to be taken into account in applying the capture data in terms of seasonal biting activity and seasonal abundance (Kettle, 1969a). In addition to those factors noted above, there were considerable day-to-day variations, mainly determined by wind and other meteorological conditions (Kettle, 1969b). There were also considerable variations in the output of adults from breeding grounds, which might or might not be reflected in the actual biting numbers. Some of these variable factors were of course measurable, such as wind speed, temperature, saturation deficiency, illumination and evaporation, providing a continuous record of data which in present times would have been admirably suited for computer analysis.

Taking all these factors into account, the records indicated that *Leptoconops becquaerti* adults were present throughout the whole year, with periods of peak abundance of August and February, and with troughs from may to June and in December. This seasonal fluctuation is strongly influenced by rainfall, which occurs in every month of the year in Jamaica with peaks in May to June and again from September to November. Superimposed on that important factor, which determines the extent of breeding areas and the output from these swamps, are tidal movements which are well known to affect the breeding grounds periodically.

In the case of *Culicoides furens*, extension of the same basic procedure revealed a very different biting cycle from that of *L. becquaerti* described above. Figure 8.2 shows the pronounced biting activity at sunset, with a steady increase in numbers up to 21.00 hours (Kettle, 1969a). The numbers decline overnight, but there is an explosive increase at dawn, reaching a peak shortly after sunrise. Some biting activity continues until after 10.00 hours, but declines thereafter.

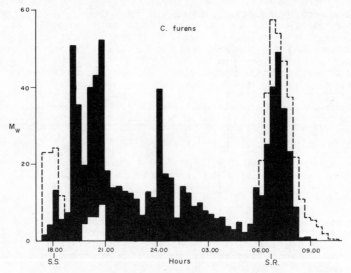

Fig. 8.2 Time of day and biting rate (M_W) of *C. furens*, Florida beach, January 1960. Black columns derived from data collected. Dotted lines and white columns have been based on standardization experiments and plotted correctly relative to sunset (SS) and sunrise (SR). (After Kettle, 1969a).

Studies on behaviour patterns of 'sandflies' in Jamaica were not completely restricted to biting activities. Observations on wind traps were also carried out to gain some idea of general flight activities, and these revealed — not altogether surprisingly — that there was not necessarily a close correlation between the two different sets of records. In the case of *L. becquaerti*, wind traps recovered maximum numbers between 12.00 and 16.00 hours in the afternoon, a time when the biting rate is low. Adults were also caught in traps at wind speeds in excess of 20 mph, whereas biting ceased at a considerably lower wind speed of 15 mph. A possible explanation is that above 15 mph the insects are unable to maintain direction and are carried along helplessly in the air currents. The two different techniques were also at variance in indicating seasonal cycles of *Culicoides*, but absolute comparison was not possible as the series of observations did not run concurrently. Some earlier work in Florida on the flight of *Culicoides furens* had already disclosed the absence of any close correlation between the numbers caught in townets, and the number biting, and it seems likely that two populations of different composition may be involved, the majority of those taken in wind traps simply dispersing rather than host seeking. The fact that these two activities have different 24-hour patterns could well be due to the fact that they are activated by different stimuli. In addition, blood feeding is a regular periodic activity in the life of the female, whereas dispersal is a much more intermittent or sporadic activity influenced by such factors as rainfall and the upsurge in emergence of adults from extensive breeding swamps.

iii. Caribbean Studies in Grand Cayman

The extension of tourism in the Caribbean, and the explosive development of hitherto neglected islands or locations, has turned the spotlight on biting midge problems very similar to those on the north coast of Jamaica. One centre of considerable entomological activity in recent years has been Grand Cayman, a small island lying between Jamaica and Cuba, about 180 miles equidistant from both. A survey using several different sampling methods showed that *Culicoides furens* was by far the most dominant species, constituting about 95% of a total catch of over 400,000 biting sand midges (Davies and Giglioli, 1977, 1979). Other species were *C. insignis*, which is widespread in distribution but low in numbers, and *C. barbosai*, which was abundant only in a few places.

In contrast to the Jamaican work, where the emphasis was predominantly on man-biting activities, the studies in Grand Cayman employed several different capture and sampling methods, including a battery-operated CDC light trap with the interesting modification that attracted midges were directed against a 3.5 inch diameter disc of mosquito screening which was lightly greased with motor vehicle lubricant in order to retain the flies, which were later examined after immersion in kerosene. In addition, a goat-baited trap and suction trap were used. Of particular interest was the extensive use of emergence traps to sample flies emerging from their various breeding grounds in mangrove and associated brackish swamp. These emergence traps had originally been tested and developed in Trinidad, and have proved a valuable tool not only for defining the distribution and relative productivity of various potential breeding areas, but also for providing valuable supplementary information regarding fluctuations in total adult biting midge population and the extent to which these are reflected in standard capture data.

In the Caymans two types of emergence trap were used: firstly, cones of tar paper covering 2.8 sq ft as used in Trinidad; and, secondly, square wooden boxes covering 4 sq ft, emerging flies being trapped on the lightly-greased inner surface of the attached collecting tube.

The great amount of information provided by these emergence traps in defining the adult output in mangrove sub-zones played a vital role in trying to unravel the complex ecology of *Culicoides*. What is particularly relevant to the present theme of this review is the light it sheds on the validity of concurrent sampling methods used to trap the population in flight. For example, the regular use of suction traps run from a mains supply showed

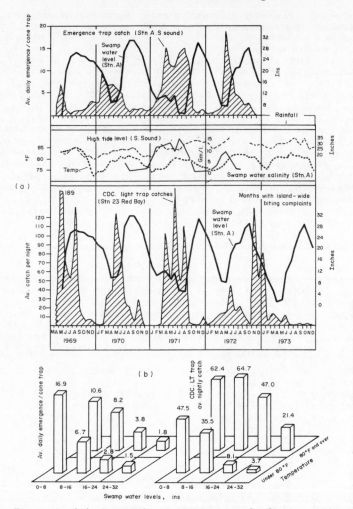

Fig. 8.3 (A) Seasonal variations in *C. furens* numbers, biting complaints, water levels and meteorological factors. (B) Trap catches at various swamp
water levels and temperatures taken from the data in
(A). (After Davies and Giglioli, 1977).

two periods of large numbers each year, with well-defined peaks from April
to June, and again in November. However, the emergence trap data indicated
a single annual peak in late spring and summer, with declining numbers
during the period October to November, when the biting nuisance was at a
peak and when high catches were also being recorded in animal-baited traps
(Figure 8.3). In addition, there were variations from year to year, making
it even more difficult to relate emergence rates directly to data obtained
from suction traps, light traps and animal-baited traps. The efficiency of
the emergence traps may of course prove to be a variable factor, influenced

in particular by rainfall, periodic floods, tidal movements, etc., making
that set of data itself difficult to standardize or interpret.

iv. The Truck Trap Technique in Florida and California

One of the most fruitful methods of studying flight patterns, not only of
Ceratopogonidae but mosquitoes as well (page 29) was the truck trap,
devised over 25 years ago in Florida and applied to great effect more
recently in California to study *Culicoides variipennis*, a biting midge widely
distributed in the US, Mexico and Canada. Flying midges are sampled by
repeatedly driving a truck trap over a course of about 3 k on a paved road,
one course in each of two study areas (Nelson and Bellamy, 1971). The
truck trap was a screened four-sided funnel of wood mounted on a pickup
truck. The mouth of the funnel, directed forwards ahead of the windshield,
was 0.6 m high and fixed 1.8 m above ground. The sides of the trap
tapered back into fine-meshed collecting bags. The truck was always driven
at a uniform rate, and headlights were only used when essential.
A series of all-night observations were carried out from May to September, one 3.2 km run being made every 15 minutes for the first six hours of
the night, and a 6.4 km run being made every 30 minutes in the latter six
hours of the night.
The results (Figure 8.4) showed peaks of flight activity by both sexes
at sunset and sunrise, the size of the peak being modified by the intensity

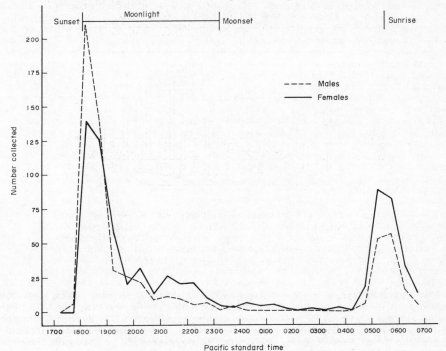

Fig. 8.4 Truck-trap collections of *Culicoides variipennis* on the night of 11/12 September 1967 (near first
quarter moon), Poso Creek study area. (After Nelson
and Bellamy, 1971).

and distribution of moonlight on different nights. A more detailed breakdown
of evening flight records was made by a series of consecutive 10-minute
collections encompassing sunset, and accompanied by continuous records of
meteorological changes within that time. On every evening of this intensive
trapping, sharp increases in the numbers of males were closely, and predict-
ably, followed by a similar increase in females.

Similar periods of intensified flight at dusk and dawn were recorded
for several other species of *Culicoides* taken in these traps. Of the possible
physical factors influencing these peaks, temperature and evaporation rates
appeared to be non-effective. Nor did peaks appear to begin in direct
response to certain specific light intensities. Although recurring peaks
sometimes began when light was the only variable undergoing change, activity was evidently not timed by light alone. Temperature itself only became
a limiting factor below 10°C in suppressing flight.

The principles underlying the use of this vehicle-mounted trap for
obtaining relatively unbiased samples of biting midges in flight, are essentially the same as those involved in another very rewarding technique,
namely suction traps. In the former the trap moves through the air, while
in the latter the air moves through the stationary trap. The truck trap
naturally samples a much greater volume of air (up to 30,000 m^3 per hour)
as compared with the much smaller volumes sucked into the suction trap
(usually less than 1000 m^3 per hour).

In the development of suction traps in general, and for *Culicoides* in
particular, British workers have been especially active, the nuisance of
biting midges being one of considerable importance in many West Highland
and Island resorts in Scotland. In one of the original experimental layouts,
suction traps in which a Vent-Axia fan sucked insects passively into a
collector, were arranged with their mouths 6 ft, 3 ft and 6 inches respectively above the ground, the catch being sorted automatically into hourly
collections. In a later series of experiments in a woodland area of southern
England, preliminary tests showed that *Culicoides* normally fly very close to
ground vegetation, suction traps with their inlets 23-48 cm from the ground
catching more adults than those at greater heights (Service, 1974). In order
to minimise the disturbance to vegetation, a height of 95 cm above the ground
was selected for comparative tests. In the experimental plan, eight traps
using Vent-Axia fans were arranged at equal distances along the perimeter of
a 15 m diameter circle. The fan in each trap delivered 634 m^3 of air per
hour into a fine wire gauze cone, and was powered by mains electricity. Of
the six different species trapped in this way, males comprised only a small
proportion. The bulk of the females (95-98%) were unfed. Only very small
proportions were freshly blood fed (0.32-1.26%), half gravid (0.17-2.21%) or
gravid (1.03-2.95%).

A further development of the suction trap has proved very useful in
studying pest *Culicoides* in Virginia (Tanner and Turner, 1975). This 'D-Vac'
is a portable knapsack type of equipment, powered by a two-cycle petrol
engine which sucks air into the end of a flexible suction hose 20 cm in
diameter. The operator walks around a prescribed 15-minute course with the
open end of the hose held about 1.5 m above the ground. This obviously has
many advantages in mobility and non-dependence on mains power supply.
Unlike the truck trap, it can be used for ambits in a wide range of environments unsuited for vehicles. In a typical field test, six collections were
made as follows: (a) one hour before sunset, (b) sunset, (c) one hour
after sunset, (d) one hour before sunrise, (e) sunrise, and (f) one hour
after sunrise. Like other sampling methods used concurrently, the D-Vac
trap effectively caught pest species — particularly *Culicoides sanguisuga* — in
large numbers, as well as capturing several species not detected by other
means.

The main object of that particular study was to compare suction trap
data with that from other trapping methods, in order to assess the seasonal

distribution of *Culicoides*.   Although the D-Vac catches were lower than by
other methods, the data followed the same seasonal pattern.   Clearly, there
are great possibilities for further developments in the use of this unbiased,
non-attractant, study technique.

vi. North American *Culicoides* Studies

Two of the most abundant biting midges along the eastern coast of
North America are *Culicoides furens* (already described in connection with the
work in Jamaica) and *C. hollensis*.   For many years these and other species
have been closely investigated by a group of research workers in North
Carolina along lines which illustrate well the development of recent trends in
flight and behaviour studies based on a variety of sampling methods (Kline
and Axtell, 1976;   Koch and Axtell, 1979a,b).   As mentioned earlier in this
chapter, the capture and sampling of *Culicoides* populations was for many
years dominated by the use of light traps in the Americas, mainly because it
was a convenient and practicable method, as well as a productive one.
In the first series of studies in coastal North Carolina in 1972-73, the
spatial and seasonal abundance of *Culicoides* was followed over a two-year
period by three concurrent techniques: light traps, sticky traps and emergence traps.
The light trap was the New Jersey type equipped with automatic timer,
and set on higher ground adjacent to the salt marsh.   Up to 32 conical
emergence traps were arranged in various vegetation zones.   The sticky
cylinder traps consisted of sections of black plastic pipe on the outside of
which a sheet of cellulose acetate, coated with tanglefoot, was wrapped.
The results showed that the light trap collections indicated the same
general seasonal trends for *C. furens* and *C. hollensis* as the emergence trap
and the sticky cylindrical trap,   *C. hollensis* being more abundant in the
spring and autumn while *C. furens* was a summer species abundant from May
through to September.
The light trap, however, did not reveal the beginning or end of the
seasonal incidence, or detailed fluctuations, as well as the other two
methods.   Emergence trap data and sticky trap data agreed fairly well with
each other except insofar as the emergence trap collected roughly equal
numbers of males and females, while few males were taken in the sticky
traps or in the light trap.
The limitations of the light trap were particularly evident in the case
of *C. hollensis* due to the fact that this species' period of greatest activity
— from two hours after sunrise until an hour after sunset — covered many
hours of daylight when the light trap was non-effective.
In a further series of investigations in coastal North Carolina in 1975
and 1976 studies were extended to the host preferences of *C. furens* and
C. hollensis.   In addition to the emergence and sticky traps described above,
animal-baited traps were employed to determine the relative attractiveness of
various domestic and wild animals (Kline and Axtell, 1976).   The animal trap
consisted essentially of a wooden frame covered with fabric screening in
which vertical slits were cut (Turner, 1972;   Tanner and Turner, 1974).   The
upper third of the trap was covered with black plastic topped by a trans-
lucent collector funnel.   Flies attracted to the caged animal inside the trap
entered through the slits and were eventually attracted by the light filtering
through the collector assembly, in which they remained trapped.
The trap arrangement on the marsh is shown in Figure 8.5 and
involved sixteen identical animal-baited traps, four emergence traps and four
sticky cylindrical traps.   The attractiveness of four domestic animals (white
rat, quail, chicken and rabbit) was first compared by randomized arrangement.   Paired comparisons were then made of the standard domestic animals

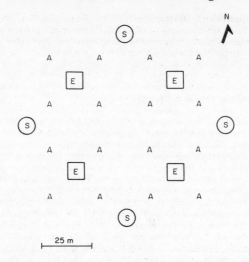

Fig. 8.5 Arrangement of animal-baited traps (A), emergence traps (E), and sticky cylinder traps (S), in a *Spartina* salt march. (After Koch and Axtell, 1979a).

with a series of local wild animals such as cotton rat, mallard duck, opossum, racoon and marsh rice rat.

The tests with the standard domestic hosts showed that the greatest numbers of both species of *Culicoides* were taken on chicken, followed in descending order by rabbit, rat and quail. The extended tests did not reveal any marked propensity for either birds or mammals as a class, and the indications are that both *C. furens* and *C. hollensis* are general or catholic feeders, with an overall lack of host specificity.

Comparisons of animal-baited traps, sticky traps and emergence traps failed to reveal any close correlation between them, and it seemed very likely that this was due to the host-seeking population being different from the recently emerged population (emergence trap) and from the unknown segment of the population trapped by the sticky cylindrical trap.

vii. Further American Studies in North Carolina

In further studies in North Carolina in 1976 (Koch and Axtell, 1979a) the flight activities of *C. furens* and *C. hollensis* were examined in more detail by the use of suction traps sited in and adjacent to a salt marsh known to be a prolific breeding ground. The objectives of this work were: firstly, to determine flight activity cycles during periods of peak seasonal abundance; secondly, to determine any correlation between flight activity and temperature, wind speed and direction; and, thirdly, to compare flight activity in an open marsh with that in an adjacent wooded area.

The suction traps used were designed in such a way that they segregated catches into hourly samples by means of jars rotating on a turntable. The mouth of the collecting funnel, into which insects were drawn by the fan current, was 1.8 m above the ground. The traps were operated at periods corresponding to peak abundance, that of *C. furens* (mainly nocturnal) being different from that of *C. hollensis* (mainly diurnal).

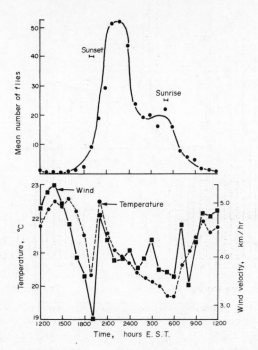

Fig. 8.6 Mean number of *Culicoides furens* collected per hour with suction traps in and near a *Spartina* salt marsh and the average hourly temperature and wind velocity.   (After Koch and Axtell, 1979b).

The results showed that with *C. furens* there are two peaks of activity during the 24-hour period (Figure 8.6), the largest peak being immediately after sunset and continuing till midnight, followed by a rapid decline in activity. A much smaller peak at sunrise is followed by a gradual decline throughout the day.

Comparison of the marsh and wooded area showed that *C. furens* remains more active by day in the latter, at least at the level sampled by the suction trap. Females formed the bulk (91%) of all catches, but the trap in the marsh area took 4–5 times as many males as in the woodland, indicating much more activity in the marsh. These comparisons could, however, have been liable to distortion if flight activity in the wooded area occurred at levels different from those in the marsh area in such a way that samples collected by the suction traps at 1.8 m were unrepresentative.

In the case of *C. hollensis* also, two peaks of activity during the 24-hour period were defined, the largest being in the morning with most activity during the daylight hours, and providing a striking contrast to *C. furens* (Figure 8.7).

Studies on the effect of wind velocity indicated a general lowering of catches when velocities exceeded 5.0 k/h, but the effect of both wind velocity and wind direction showed significant differences in effect on *Culicoides* in the woodland as compared with the marsh environment.

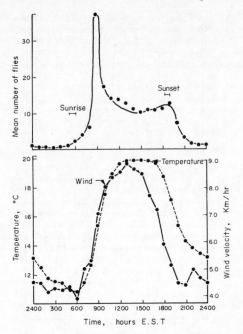

Fig. 8.7   Mean number of *Culicoides hollensis* collected
per hour with suction traps in and near a *Spartina*
salt marsh and the average hourly temperature and
wind velocity.   (After Koch and Axtell, 1979b).

viii. *Culicoides* as Vectors of Animal Disease in Africa

     None of the studies discussed so far have been concerned with biting
midges as vectors of disease, and they have also been mainly restricted to
defining seasonal and diel activity patterns, and biting activity on man.
     In those areas where *Culicoides* are established vectors of viral
disease, particularly of domestic animals, such as Blue Tongue of sheep and
African horse sickness, studies on the vectors have had to be extended to
such topics as vector status, longevity and host preference.   Work along
these lines is being actively pursued in several African countries, such as
Kenya, Uganda and Nigeria, and has provided new information about the
behaviour patterns of biting midges.
     Intensified interest in arboviral disease, and other vector-borne disease
in Panama connected with great environmental changes attending the construction and completion of the Bayano River Dam and hydroelectric scheme,
have disclosed a large number of isolations of unidentified Simbu-group
viruses from *Culicoides*.   This finding emphasizes the potential importance of
this group of insects and the need for more detailed studies on its ecology.
     The intensified work in Kenya in the last few years has been concerned
firstly with determining which of the many suspect *Culicoides* are involved in
the transmission of Blue Tongue in sheep, and ephemeral fever of cattle
(Walker and Boreham, 1976; Walker, 1977).   Apart from routine virus isolation
from wild-caught females, high priority has been given to defining the
seasonal distribution and the blood feeding habits.   As in so many recent

biting midge studies, the battery-operated light trap was found to be the simplest technique for obtaining large numbers of adults, even though the interpretation of light trap data in terms of normal behaviour patterns is a difficult and controversial one.   Light trap methods were supplemented by the use of non-attractant battery-operated suction traps.

As longevity is an important factor determining vector potential, the samples of females caught were age-graded into six categories: nulliparous unfed, nulliparous blood-engorged, pre-gravid, fully gravid, parous unfed and parous blood-engorged.   Unfortunately, the advanced technique for determining 'multiparity' (i.e. the number of complete gonotrophic cycles undergone by each female) could not be done in the way in which it has been so successfully applied to mosquitoes.   However, the proportion of nulliparous to parous flies could be used to calculate longevity and to indicate seasonal variations in reproduction.   Conclusions from such calculations, of course, necessarily involve several assumptions, and perhaps too much should not be read into these theoretical conclusions.

The siting of the light traps and suction traps near sheep and cattle pens meant that attraction to the animals in these pens could largely determine the composition of the *Culicoides* population.   Because of the presence of these host animals, a much higher proportion of freshly-engorged females was taken than has been the experience with similar traps in other countries. This made it possible to carry out a good series (682) of precipitin tests on blood meal samples, the bulk of these from the two dominant species, *C. pallidipennis* and *C. schultzei*.   Not unexpectedly, they showed that blood feeding was predominantly bovid and sheep/goats.   Traps operated away from livestock pens produced very few engorged specimens, indicating that engorged females probably seek resting sites in the immediate vicinity of the feeding site.

It is interesting to note that, in contrast to so many of the study centres on the Atlantic and Gulf coasts of the United States, and in the Caribbean, man-biting *Culicoides* were rarely encountered in this particular study and are apparently rare in Kenya in general.

CHAPTER 9

Phlebotomine Sandflies

i. Introduction

The problems posed in the study of *Culicoides* and its biting midge
allies are shared in many respects by another group of small, elusive,
blood-sucking flies, namely the phlebotomines, or sandflies proper. Work on
Phlebotomus, however, in contrast to that on *Culicoides*, has been motivated
not by any nuisance or pest status, but by its important role as a vector of
human and animal leishmaniases. The widespread distribution of leishmaniasis
has meant that studies on the vector *Phlebotomus* have by no means been
restricted to the tropics and sub-tropics, but have been pursued in countries

of the temperate zone, particularly the USSR and the Mediterranean countries (Killick-Kendrick, 1978; WHO, 1979). In recent years one of the most productive study centres has been in the south of France, where there has been a new and vigorous approach to basic problems in the ecology and behaviour of *Phlebotomus*.

Perhaps the best way of demonstrating the range of problems encountered, and the different ways in which these have been tackled by different research teams, is to use as a model the research carried out in recent years in central and south America by teams in Belize (formerly British Honduras), Brazil and Panama. In all of these areas human leishmaniasis is a zoonosis, i.e. a disease normally endemic in certain wild animal populations, and in which man is the occasional or accidental host, but nevertheless a very vulnerable one.

In the preliminary stages leading up to more specialised vector studies, the ground work is laid by parasitological studies on the human population and on a wide range of suspect animal reservoirs of the disease, mainly rodents and other small mammals, in order to determine the incidence and distribution of leishmania lesions and infections. Such surveys usually narrow the range of suspect animals to a few main reservoir species. From then on, the main entomological problem is to define the hosts of likely vector species — in this case *Phlebotomus* — firstly with regard to transmission of the parasites between animal hosts, and secondly with regard to transmission from animal to man. This is a basic problem common to all three study areas, but in Panama *Phlebotomus* is also important in its role as vector of bartonellosis and of a particular virus associated with vesicular stomatitis (Gorgas Memorial Laboratory, 1977; Aitken and co-workers, 1975).

While all three groups of entomologists involved were facing common problems of establishing valid sampling techniques for *Phlebotomus*, defining seasonal incidence, host specificity, etc., the two British-dominated groups in Belize and in Belem, Brazil, used different research methods and a rather different approach from their American colleagues based in the Gorgas Memorial Laboratory and associated US research units in Panama. The outcome of this slightly divergent plan of attack has been advantageous in that the gaps or shortcoming in one area   have in many cases been compensated by the successes or progress achieved by the other.

ii.   *Phlebotomus* Investigations in Belize

Following the confirmation that dermal leishmaniasis in British Honduras was clearly a zoonosis with several species of jungle rodents incriminated as reservoir hosts, it became imperative to find out what species of *Phlebotomus* normally feed on those animals and could act as vectors (Disney, 1968). The problems of devising satisfactory techniques for this purpose is not an easy one. Many entomologists elsewhere, including Panama, had tried out various designs of rodent-baited trap, but in most cases the catch was disappointingly low. One of those methods involved direct catching on caged animals, which were visited by a collector at half-hourly intervals; greater numbers of flies were obtained this way, but these appeared to be mainly man-biting species (Thatcher and Hertig, 1966). It was therefore essential that any technique should trap efficiently and mechanically only those species attracted to a specific bait animal in the complete absence of man.

The method finally devised in Belize after much trial and error owes its effectiveness to the use of a castor oil film surrounding the bait animal cage. This film effectively caught and retained flies approaching the bait with their usual short hopping flight, and maintained them in good condition for later examination (Disney, 1966). Details of the trap are shown in Figure 9.1. The normal checks were then carried out to ensure that flies

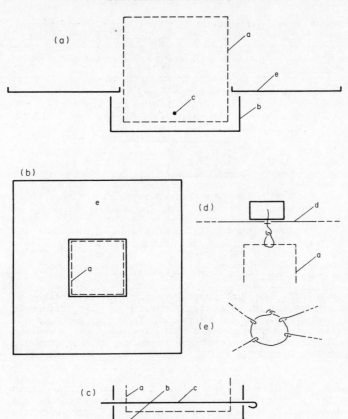

Fig. 9.1 Diagrams of trap and combined roof and trap suspension. (A) Trap in vertical section; (B) trap in horizontal section at level of catching surface (at half the scale of upper diagram; (C) cage tray and lower part of cage in section, showing securing wire; (D) centre of roof in vertical section showing trap support; (E) central part of roof suspension: (a) rat cage; (b) cage tray; (c) securing wire; (d) roof; (e) catching surface. (After Disney, 1966).

attracted to the trap were in fact attracted to the animal bait and not to the oil film; this involved a series of staggered tests with alternating absence and presence of bait. When the attraction of the bait had been clearly confirmed, a preliminary series of tests were carried out as follows.

With the catching surface arranged 110 cm from the ground, four species of rodent were used as bait, all of which were incriminated at that time or subsequently as reservoirs of leishmania. Out of seven species of *Phlebotomus* taken, one (*Lutzomyia olmeca*) was attracted to the rodent-baited traps in greatest numbers, and appeared to be the most likely vector in the rodent-to-rodent cycle.

More detailed studies on host preference were carried out by erecting

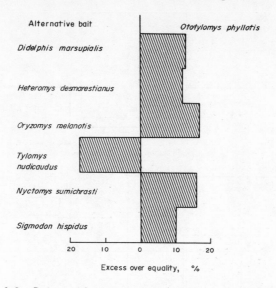

Fig. 9.2   Bait preference trials.   The percentages of
Lutzomyia flaviscutellata   attracted   to   *Ototylomys
phyllotis* and to an alternative bait.   To the right the
percentages in excess of 50% attracted to *O. phyllotis*.
To the left the percentages in excess of 50% attracted
to the alternative bait.   (After Disney, 1968).

traps in pairs with their centres about 2 m apart.   One of the pair was
baited with the rodent *Ototylomys phyllotis* as standard, and the other with
the alternative bait, one animal to each trap.   Traps were changed round
each day to eliminate position effect.   The results of 110 trapping nights
extending over six months are depicted in Figure 9.2, with special reference
to *Lutzomyia olmeca*, and show that this species has a distinct preference for
the important leishmania host *Ototylomys* over all the other baits offered,
with the exception of the more strongly attractive *Tylomys*.
    Traps baited with *Ototylomys*, set up at different heights above ground
level from 0.1 m to 11–12 m, showed a clear vertical zonation of biting flies,
with the significant finding that the preferred biting zone of the proven
vector, *L. olmeca*, which was below 4 m, was the only zone in which all
known animal hosts of leishmania had been trapped.
    The problem of vertical zonation of biting was approached from another
angle, bearing in mind that in Belize man can become infected with leishmania when working on the forest floor, even though no *Phlebotomus* caught
biting man in the forest had been found infected with the parasite.   This
suggested the possibility that the vector of leishmania to man could be a
Phlebotomus which was normally arboreal in habit (Williams, 1965).   To
examine this, a wooden tower with two platforms was set up, enabling simultaneous collections to be made at ground level and at 25 ft and 40 ft above
the ground (Williams, 1970).   The bait in this experiment was human, with
one man acting as bait and one as collector at each level, the roles being
reversed after two hours.   After a further two hours the men moved to
different levels.   The three collecting periods selected covering 24 hours
were from 06.00 to 14.00, 14.00 to 22.00 and 22.00 to 06.00 hours.   Over the
same period rodent-baited oil traps were set up on one day and one night
each week.

Nineteen species of *Phlebotomus* were taken in this series, eleven of these on man. *Lutzomyia cruciata* accounted for more than 50% of the man-biting catch at each level. Sometimes it was the commonest man biter at all heights, while at other periods two other species, *L. panamensis* and *L. shannoni*, were more abundant. At ground level *L. panamensis* showed consistently the greatest biting density. Once again it was evident that the only proven vector of leishmania, *Lutzomyia olmeca*, is not attracted to man and rarely taken on human bait. The question still remains, therefore, about the exact identity of the *Phlebotomus* vector transmitting leishmania from rodent to man.

iii. Investigations in Belem (Brazil)

With the experience gained in Belize, more intensive parasite studies were initiated in Belem, Brazil, in 1963, where, despite the great deal of work which had been done over the years, the exact composition of the forest fauna providing a reservoir of disease — almost certainly rodents — had still to be clearly defined. This first exploratory trip led to the finding of the first particular leishmania lesions previously disclosed in Belize, in a small forest rodent *Oryzomys goeldii*, captured in the Utinga forest about 4 km from Belem. Subsequently a large number of infected *Oryzomys* were found, and a groundwork of seasonal incidence together with the more detailed knowledge of the parasite was established.

The whole situation came under intensive investigation by the British team from 1965 onwards, with the main object — as in Belize — of incriminating the *Phlebotomus* vectors involved in the rat-to-rat and rat-to-human cycles (Lainson and Shaw, 1968; Lainson and co-workers, 1973; Shaw and Lainson, 1968, 1972; Shaw and co-workers, 1972; Ward and co-workers, 1973). The two most important rodent hosts to emerge were *Oryzomys* and *Proechimys* and these were used in tests based on precisely the same 'Disney' trap which had proved so successful in Belize, the only modification being an increase to approximately twice the size to accommodate certain larger animals as bait. As these two rodents are ground dwellers, seldom climbing any distance, traps were set at ground level and operated overnight.

All the rodents used as bait were found to attract predominantly one species of phlebotomine, namely *Lutzomyia flaviscutellata*. Table 9.1 shows four months' records of catches, and also demonstrates that this species rarely bites man (Lainson and Shaw, 1968). Of the two rodent species originally tested, one — which was also a natural host of leishmania — appeared to be the most attractive bait for *L. flaviscutellata*; this was *Proechimys guyannensis*, and this was the one selected for quantitative studies using Disney traps. Traps set up at ground level in different locations revealed not only considerable nightly variations but also showed that some sites consistently produced higher catches than others.

Table 9.1 Biting habits of *Lutzomyia flaviscutellata* in relation to rodents and man, Utinga Forest, Belem, Brazil, June–October 1967. (After Lainson & Shaw, 1968).

Bait	Total number of phlebotomines captured	Number of *L. flaviscutellata*	Percentage of *L. flaviscutellata* in total catch
Man	724	5	0.69
Rodent	2,774	2,731	99.15

Catches from rat-baited traps were compared to catches on man, and again showed the overwhelming predominance of *L. flaviscutellata* on all baits except man. While female *Phlebotomus* predominated in these catches, males were also taken, occasionally in high numbers. As the bait itself appeared unlikely to attract the non-bloodsucking male, their presence probably reflects an attraction to the aggregation of females in the vicinity of the bait.

At this stage it was becoming clear that at least two species of *Leishmania* were responsible for cutaneous leishmaniasis in north and central Brazil, and this led to a complete revision of classification of those parasites. These parasitological studies in turn stimulated further entomological work in northern Brazil, and it was found that the most important vector of the disease in man — as distinct from rodents — was *Psychodopygus wellcomei* (Ward and co-workers, 1973). This species was found to bite man throughout the day, even at high light intensities, that is at the very time when men are likely to be working in the forest. Forest clearing for the construction of new roads and the development of mining areas in Brazil provide situations where labour forces are exposed to increased risk of *Leishmania* infection from *Phlebotomus* bites.

As well as being an avid man-biter, *Psychodopygus wellcomei* was also shown to be strongly attracted to rodents and would therefore appear to be the natural vector of leishmaniasis from rodents to man, in contrast to *Lutzomyia flaviscutellata* whose preference for rodents makes it the likely main vector in the rodent-to-rodent cycle. The distribution of bites by these and other species are illustrated diagrammatically in Figure 9.3.

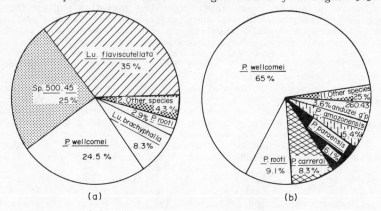

(a)   (b)

Fig. 9.3 (A) Rodent-biting phlebotomine sandflies captured on Disney traps, and (B) Man-biting sandflies captured, in the Serra dos Carajas, Para, Brazil. (After Ward and co-workers, 1973).

While the biting habits of potential vector species was the main theme of these investigations, other aspects of vector behaviour were also examined, particularly the nature of the normal outdoor resting places. In this, however, considerable effort was less well rewarded, and the findings were virtually negative. In Belize a paucity of anthropophilic *Phlebotomus* had also been noted in extensive searches in tree buttresses and similar places. As will be noted later in this chapter, this puzzling experience in forest areas where vast numbers of sandflies are known to exist is in sharp contrast to experience in Panama where the detection of outdoor resting sites, and the collection of large numbers of flies, has played a prominent part in vector studies.

In the Brazilian work based on Belem an extensive trapping programme, using Disney traps baited with spiny rats (*Proechimys guyannensis*), was set up to determine the seasonal abundance of sandflies in two different types of forest (Shaw and Lainson, 1972). In both types seasonal fluctuations in density followed a similar pattern in relation to rainfall, numbers of *L. flaviscutellata* falling off with the onset of the heavy rainy season (Figures 9.4, 9.5). It should be noted that these figures for seasonal abundance of flies were based entirely on routine captures in rodent-baited traps and that, in contrast to the experiences in Panama, trials with other supplementary methods of sampling such as light traps and outdoor collections did not produce anything of value.

In further studies on feeding behaviour of *L. flaviscutellata*, with particular reference to biting habits at different heights, rodent-baited Disney traps containing spiny rats were set up at heights of 0.2, 5.0, 10.0 and 15.0 m in the Utinga forest, and the results compared with man-baited catches in the same forest carried out between 21.00 and 24.00 hours (Shaw and co-workers, 1972). Previous catches had already indicated that *L. flaviscutellata* was predominantly a ground-level feeder, constituting 98% of the flies attracted to small rodents at that level. This had also been the experience with the closely allied *L. olmeca* in Belize, which was regarded as a lower zone species with a ceiling around 8 m from the ground.

The series of catches at different levels provided ample information of lower zone activity, showing a marked fall-off in numbers of flies caught above 0.2 m, indicating that this species really is very restricted — in biting activities at least — to quite a narrow zone near the ground. These findings also helped to explain why such typically arboreal forest animals as opossums and monkeys, living at a level well above that of *Lutzomyia*, show low to negative incidence of leishmaniasis, even though these flies are attracted to and feed on these animals in traps at ground level.

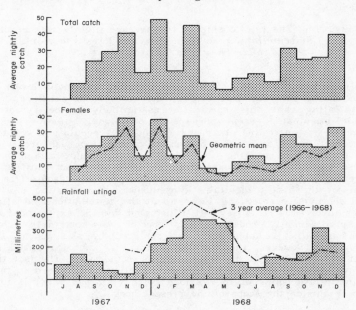

Fig. 9.4 The relationship of the average nightly catches of *Lu. flaviscutellata*, caught in the Utinga forest, to rainfall. (After Shaw and Lainson, 1972).

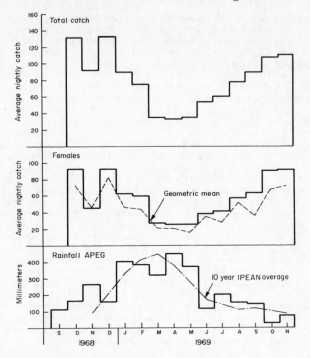

Fig. 9.5 The relationship of the average nightly catches of
Lu. flaviscutellata, caught in the Catu forest, to rainfall.
(After Shaw and Lainson, 1972).

The man-baited catches, sitting and standing at ground level, produced
a very low biting rate for *Lutzomyia* of the order of only 1-2 per man hour,
further indicating that this species seems unlikely to be a rodent-to-man
vector of leishmaniasis, except perhaps in occasional situations such as in
the wetter forests where the increased *Phlebotomus* population would increase
man/fly contact in the rare event of humans penetrating such areas.

The problem of studying *Phlebotomus* ecology and behaviour patterns
under these varied conditions has been rendered even more difficult by the
need to review the taxonomy and nomenclature of this group from time to
time. This is the almost inevitable consequence of penetrating new fields of
study and continually enlarging ideas about the distribution and affinities of
the species involved. This is well illustrated by the experience in Belize,
where the predominantly rodent biting species was originally classed as
Phlebotomus apicalis.   Later this was replaced by a synonym, *Lutzomyia
flaviscutellata*. However, that species in turn is now regarded as a species
complex of which the Belize representative is *L. olmeca*. The fact that these
studies have revealed an equally complex situation with regard to the classification and relationships of the various *Leishmania* parasites, is fortunately
one which is outside the scope of the present book.

iv. Studies in Panama

The Gorgas Memorial Institute of Tropical and Preventive Medicine is one of the longest established research centres in tropical America (Gorgas Memorial Laboratory, 1977). For over 50 years, since its establishment in 1929, it has been a working base for a succession of outstanding biologists and medical men covering all aspects of tropical disease, particularly insect-borne diseases and their associated vectors. Among the many problems has been that of *Phlebotomus* and *Leishmania* transmission, presenting a similar challenge to that already described for Belize and Brazil. In addition, the question of *Phlebotomus* as a vector of vesicular stomatitis virus and other viral infections has drawn increasing attention in recent years. Each of these two aspects of *Phlebotomus* as a disease vector has tended to attract the attention of different groups of research workers even within Panama, and this has provided stimulating variations in the approach to the essential central problem — that of the ecology and behaviour of *Phlebotomus*.

Intensive studies on the natural hosts of *Phlebotomus* in relation to the likely reservoirs of cutaneous leishmaniasis had started in 1956, but the special studies selected for attention in the present context were carried out in 1962 and 1963, thus pre-dating the intensive work in Belize from 1964 onwards. Attempts to devise a trap for *Phlebotomus* which could be baited with live animals and left unattended in the forest were less successful than the later work in Belize; various designs which had proved useful in other situations elsewhere gave disappointing results. The bait animals, suspect reservoir hosts of *Leishmania*, included animals such as the opossum (*Didelphys* and *Caluromys*), sloths (*Bradypus*) and kinkajou (*Potos*), which were considerably larger than the forest rodents studies in Belize and there-fore required correspondingly larger cages.

Among the various possible techniques tested, and rejected, was a trap baited with small animals and surrounded by oiled paper frames hung around the sides and open ends. One of the difficulties was that in the very humid climate of Panama the castor oil arrestant, which had proved successful in drier countries, tended to run off leaving the paper with insufficient adhesive.

The Panama team then decided they would have to fall back on the practice of direct collecting from bait (Thatcher and Hertig, 1966). Animals in cages were set up in two or more groups on pole racks about 4 ft above the ground. Sandflies had access to the bait through the bottom of the cages, as well as from all other directions. The animals were placed in position before subset, then the collectors retired to an adjacent cleared area, 50-100 yards away. From 19.00 hours all bait animals were checked at half-hourly intervals, and sandflies found on the animals were collected by suction vials. Many of the flies caught had already blood-fed, while most of the others were attempting to feed. The actual time spent near the animals by the collectors amounts to about five minutes every half-hour, thus reducing the human influence to a minimum. Panama has at least 65 species of *Phlebotomus*, and many of these were taken in direct bait collections. The trials showed that several of the dominant man-biting species in the area were taken in large numbers on animal baits, demonstrating a wide host rage, with a possible preference for the kinkajou (*Potos*) over other available animals.

In 1965 the advantages of the Disney trap developed in Belize were discussed during a visit to Panama by one of the British team, and as a result extensive trials were carried out with this trapping technique on Panamanian *Phlebotomus* over the next two years. The basic design was modified slightly; instead of the single oil-containing pan, or moat, surrounding the bait animal, up to three pans were used, two being the most usual number. The additional pan was situated below the level of the animal. The moat pans were larger than those used in Belize — 66 x 94 cm as compared to $7\frac{1}{2}$ x $7\frac{1}{2}$ ins (19 x 19 cm) — and the other dimensions were correspondingly larger to accommodate the larger bait animals (Thatcher, 1968).

Using the common opossum as bait, initial trials at ground level showed that both *P. trapidoi* and the man-biting *P. panamensis* were readily attracted to the opossum bait. This was followed by a series of comparison trap catches at ground level (1.2 m) and at canopy level (10–13 m) using three bait animals, viz. opossum (*Didelphys*), kinkajou (*Potos*) and porcupine (*Coendou*).

In 82 trap nights over 7,000 sandflies were caught, including 630 fed females. Of the three dominant species, 95% of *P. panamensis* were taken in the low trap, while 97% of *P. trapidoi* and 90% of *P. sanguinarius* were taken from the high trap. On several individual nights there was complete separation between *P. panamensis* in the low trap and *P. trapidoi* in the high trap. All three of these species are common man-biters at ground level.

v.   From 1969–70 onwards, intensive studies on *Phlebotomus*, mainly in its role as virus vector, were carried out by another American team in Panama in a series of outstanding studies dealing with activities and behaviour patterns of sandflies in all aspects — feeding, resting and flight movement (Chaniotis and co-workers, 1971a,b, 1972, 1974; Chaniotis and Correa, 1974; Rutledge and Ellenwood, 1975; Tesh and co-workers, 1971). One of the principle objectives was to determine the relative seasonal abundance of *Phlebotomus*, utilizing several sampling methods. Again, various types of animal-baited, CO_2 traps and oil traps were tested and rejected. Two methods were finally selected for intensive studies, namely light traps in the form of small portable battery-operated models for sampling the actively flying population, and aspirators for sampling the resting population. In each locality two light traps were operated, one near the ground and the other in the forest canopy — 14 to 25 m according to the type of vegetation — for two successive nights each week. Resting adults in various types of vegetation and undergrowth (tree trunks, etc.) were captured with conventional aspirator with the aid of a flash light.

During the 59-week study period a total of 60,455 sandflies, with approximately equal sex ratio and representing 37 species, were captured. A selection of different patterns of capture is shown in Table 9.2, from which it can be seen that some species (1, 2 and 4) were attracted almost exclusively to light. Others (e.g. 7) were taken equally at light and in resting collections. In another group (3, 5, 6 and 8) resting captures greatly exceeded light trap catches. In comparison between ground and canopy

Table 9.2 Number of different species of sandflies taken in Panama by three concurrent catching/sampling techniques. (After Chaniotis, 1971).

Species	Light trap Ground level	Canopy	Aspirator
1. *Lutzomyia aclydifera*	2675	513	3
2. *L. carpenteri*	2303	415	2
3. *L. ovalessi*	71	23	551
4. *L. panamensis*	5152	1615	43
5. *L. roratensis*	7	4	632
6. *L. shannoni*	128	44	9799
7. *L. trapidoi*	2092	3128	2026
8. *L. trinidadensis*	233	85	12687

light trap catches, 1, 2 and 4 had greater ground catches, while 7 had greater catches in the canopy. The main man-biting species, *P. panamensis*, was one of those taken predominantly in light trap catches, mainly at ground level, and was comparatively rare in resting collections.

Attention was then turned to more detailed observations on the important man-biting species with regard to daily and seasonal activities in a tropical forest. Collections on human bait were made for one hour each at four periods staggered in such a way as to fully represent the periods of noon, dusk, midnight and dawn. Weekly schedules ran alternately between ground level and a canopy platform 28 ft above ground. Of the 37 species recorded in this field station, seven were regularly attracted to and landed on the human bait. It is interesting to note that none of the five species of *Phlebotomus* which had been recorded on human hosts under similar conditions in Belize were taken in these bait collections even though they were present in the Panama study area. In addition to the expected females, males were recorded both at ground level and in the canopy. The types of biting activity at both levels are shown in Table 9.3. Most striking are the two contrasting behaviour patterns exhibited by man-biting *Phlebotomus*, with one group including *L. olmeca* and *P. panamensis* dominant at ground level, and the other group typified by *L. trapidoi* dominant in the canopy.

Table 9.3 Vertical stratification of man-biting activity of seven anthrophilic species of phlebotomine in Panama (after Chaniotis, 1971).

Species	Percentage of catch		Total numbers
	Ground	Canopy	
Lutzomyia gomezi	5.6	94.4	71
L. olmeca	91.3	8.7	46
L. panamensis	90.7	9.3	420
L. pessoana	86.4	13.6	1412
L. sanguinaria	27.6	72.4	214
L. trapidoi	6.8	93.2	1331
L. ylephiletrix	51.2	48.8	3533

With regard to the daily cycle of biting activity, the results corroborated experience elsewhere that phlebotomines are primarily crepuscular and nocturnal biters. Seasonal activity of man-biting species showed them to be active throughout the year at both ground and canopy levels. However, when these figures were compared with seasonal population densities in the same area recorded over the same period but based on different sampling techniques (light traps and resting catches) there appeared to be no consistent relationship, and indeed some months showed wide divergences (Figure 9.6).

From this it appeared that the degree of biting activity was determined not only by population density but by other factors as well. There were times, for example, when biting activity was very low even though man-biting species were abundant near the collector.

Comparisons between biting activity of anthropophilic species — based on human bait collections — and flying activities as judged by light trap catches, were made in two different sites, namely mature forest and an adjacent open space 50 m away. These figures showed that flying activity in the open space was about one-fifth of that in the forest, while biting was one-fifteenth of that in the forest over the same period.

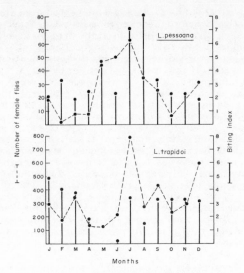

Fig. 9.6 Comparison of monthly biting activity and density of population for *L. pessoana* and *L. trapidoi*, the most abundant anthropophilic species in the Limbo Field Station at ground level and in the forest canopy, respectively. The density curve was based on data obtained from another project which was run concurrently. (After Chaniotis and co-workers, 1971a).

vi. Of the various sampling and study methods used in the Panama work, one of the most productive was based on intensive investigations into the natural outdoor resting places of *Phlebotomus*. All those working with forest phlebotomines have been aware that high populations of flies must exist in daytime shelters among trees or other vegetation, and that such populations must include females in all the various stages not immediately concerned with host-seeking and blood-feeding. This would include, for example, all recently emerged females not old enough to take a blood meal, and all females which had gorged and in which the blood meal was digesting and the ovaries maturing. As has been seen, sustained efforts to detect this resting population in the Brazil studies were almost completely unsuccessful. In Belize, initial searches on tree trunks, crevices, etc., yielded several species, but not those of the main rodent and man biters. However, the resting places that did produce the most *Phlebotomus* in that work were under dead leaves on the forest floor, where the catches were dominated by *L. olmeca* and *L. panamensis*. The catches were never high, a total of 73 specimens of the former and 147 of the latter being taken in weekly collections over a period of nine months (Disney, 1968).

In contrast to this, high populations were revealed in the intensive searches in Panama carried out over two years in every possible resting site (Chaniotis and co-workers, 1972). Over 7,000 flies (males and females) were recorded from nine different diurnal resting sites. These high catches opened up several lines of investigation not otherwise possible. For example, routine collections provided a separate assessment, as distinct from light trap data, on which to plot seasonal changes in population density. It was also possible to make valid comparisons between the attractiveness of different resting sites, and also between resting sites at different heights from ground level upwards. This opened up the possibility of studying actual movements of flies, horizontal and vertical, between these different sites. Outdoor collections also provided information about females in a wider spectrum of physiological stages than those caught by bait alone. Among these physiological grades were engorged females taken in numbers sufficiently high to enable a precipitin test survey of blood meal origins to be carried out.

With regard to the first line of investigation, intensive searches which disclosed high populations of resting *Phlebotomus* now made it possible to define more accurately the preferred diurnal resting sites of a number of different species in the forest. Among the places searches were animal burrows, tree hollows, leaf litter, green plants, three trunks and buttresses at two levels, viz. from ground to 0.6 m and from 0.6 m to 2 m, and finally tree trunks at three levels, namely 5, 9 and 15 m above the ground. These findings showed that different species tended to dominate different microhabitats, and this applies particularly to sites on tree trunks at different heights above ground (Table 9.4).

Significant seasonal changes in species composition were recorded, particularly in tree hollows and animal burrows, some species being dominant in the rainy season and others in the dry season (Table 9.5). Reference to the table shows that in the wet season the population was dominated by species J, I and D. In the dry season dominant species were J, H, K and A. Species A and F were totally absent in the wet season.

Table 9.4 Daytime resting places of sandflies in Panama.
Vertical distribution of dominant species on tree trunks.
(After Chaniotis et al., 1972).

Height	Species
0.0 to 0.6 m	*L. trinidadensis* and *L. shannoni*
0.6 to 2.0 m	*L. trinidadensis* and *L. trapidoi*
5 m	*L. trinidadensis* and *L. rorataensis*
9 to 15 m	*L. rorataensis* and *L. micropyga*

Table 9.5 Seasonal changes in sandfly composition in samples from tree hollows in Panama. (After Chaniotis et al., 1972).

Species	Percentage of total catch (all species) Dry season	Wet season
A. *Lutzomyia composoi*	10.1	0.0
B. *L. dasymera*	1.3	3.0
C. *L. hansoni*	1.5	2.0
D. *L. nordesina*	2.1	6.4
E. *L. pessoana*	2.7	0.6
F. *L. saulensis*	1.6	0.0
G. *L. shannoni*	2.4	1.7
H. *L. trinidadensis*	16.0	3.9
I. *L. vesicifera*	4.1	20.6
J. *L. vespertilionis*	42.6	54.1
K. *Warileya nigrosacculus*	11.6	3.9
Total number (all species)	751	534

The anthropophilic species tended to dominate one particular type of site, namely green plants, shrubs and saplings in the more open parts of the forest floor. Of the high catches recorded there (2387 flies) *L. pessoana* made up 96%.

The availability of large numbers of *Phlebotomus* captured from diurnal resting sites made it possible to plan a marking-release-recapture experiment in an attempt to define the horizontal and vertical movements of sandflies in a Panamanian rain forest (Chaniotis and co-workers, 1974). Flies captured on tree trunks during the day were released at the base of a large tree in the centre of the experimental area. Additional material for marking and release was provided by flies caught at platform level after dusk, 95% of the catch being *L. trapidoi*, and subsequently used in platform release experiments.

Over three years nearly 8200 flies were captured, marked and released at ground level on three different occasions. Encouragingly high recapture rates, varying from 2.6% to 24.3%, were recorded. Similar large releases of over 11,000 marked flies were made at canopy level 30 m above the ground. The numbers recaptured were sufficient to show the limited range of flight of phlebotamines, and their tendency to remain localized.

Flies are totally absent from canopy platform levels during daylight hours, the preferred diurnal resting site of *L. trapidoi* being forest leaf litter. The recapture of 119 specimens on the platform at night demonstrated the tendency of the flies to return to the canopy from adjacent resting sites on the lower part of tree trunks and on forest litter.

With regard to the other avenues of research opened up by the finding of large diurnal resting populations of flies, including many with blood meals, it was now possible to carry out an extensive precipitin test series on samples of the sandfly population which were not obviously biased by the presence of any particular bait animal (Tesh and co-workers, 1971). In studying this aspect of the behaviour of forest phlebotamines, two problems arose. First was the relatively small volume of blood ingested by sandflies, which is only about one-tenth of the amount taken in by the average mosquito feed and therefore limits the number of tests which can be done on each specimen. To overcome this difficulty a microplate modification of the tube precipitin test was developed, which enabled smaller volumes to be used for several tests. The second problem was the tremendous diversity of potential hosts in such a tropical rain forest — in the case of Panama over 200 mammal, nearly 700 avian and 100 different reptile and amphibian species.

Sandfly blood meals were tested in two stages. In the first, diluted blood meals were screened against three broadly reacting class antisera (mammal, bird or reptile-amphibian). If adequate numbers of specimens were available, those reacting to one of these three, say mammal, were then tested against six order-specific antisera, viz. primate, edentate, rodent, marsupial, carnivore and cheiropteran. At this stage distinct host preferences began to appear. *L. trapidoi*, for example, established as an important anthropophilic species on the basis of human bait collections, showed the broadest range of blood feeding with 21% feeding on primates, 28% on edentates, 30% on rodents, 10% on marsupials and 10% on carnivores. *L. trapidoi* also recorded the highest number of primate feeds. Of three other abundant species, the first showed 76% feeding on edentates, the second showed 45% of feeds on rodents, while the third recorded 100% reacting to bat antisera.

The edentate-positive bloods could be further differentiated into anti-sloth and anti-armadillo. With regard to primate positives, these could have referred to any one of the five known species of primate inhabiting the forest, or possibly to the occasional man entering that area.

Because of the richness and variety of the potential host fauna, and likely fluctuations in their numbers from place to place and from time to time, it was very difficult to form a clear idea of the relative population of potential hosts present in each site, and quite impossible to arrive at any figure for the Forage Ratio (see page 58). Local differences in the feeding pattern of the same species of sandfly in different localities may well have been due to changes in host fauna, the temporary absence of the preferred animal possibly inducing feeding on other less-favoured hosts still present.

vii. European Studies on *Phlebotomus*

Intensive studies have been carried out by Russian workers on the phlebotomine vectors of leishmaniasis since the early 'thirties, particularly in relation to the main natural animal reservoir of the disease, the great gerbil (*Rhombomys opimus*). In this and in other vector studies, Russian work has continually emphasized that a thorough knowledge of the vector, and in particular of its density, is essential for a better understanding of the epidemiology and epizootiological situation prevailing in natural foci of disease. With regard to *Phlebotomus*, they have pointed out that no standardized method exists for determining the density of sandflies, and that consequently the data derived from different investigators are often not comparable and do not always reflect the real situation. They have found that none of the methods applied so far to determine fly density or other various indices lend themselves to statistical analysis. Many of the methods used with other related groups of Diptera have not been found applicable to *Phlebotomus*, and it was considered that such widely used techniques as light traps and the use of different animal baits may provide a false picture of the species composition and density of a sandfly population. In view of the fact that different species respond in different ways to artificial light, and even more to choice of blood meal source, they have concluded that the most valid and unbiased sampling method is the use of standardized sheets of sticky paper (20 x 30 cm) covered with castor oil (Dergacheva and co-workers, 1979). They were the pioneers of this particular technique, which has now become so widely used in many other countries.

In areas occupied by animal colonies, particularly of the great gerbil, captures of sandflies are carried out on suspended arched sheets of sticky paper, each sheet being fixed on a peg at a height of 4-5 cm above the ground, with the convex side towards the ground. The sandfly density in burrow colonies is determined by pegging ten arched papers distributed over the surface of each colony, and leaving them overnight. In order to standardize such catches, papers are set at equal distances from various burrow entrances, with ten different colonies being sampled. Pegged sheets of paper are placed over both inhabited and uninhabited burrows of the great gerbil, as well as in the habitats of other animals. The data from sticky paper captures is considered to be subject to much less variability than other capture methods, and consequently more acceptable for statistical analysis.

The same standardization is aimed at when phlebotomines are sampled in human settlements and all types of associated farm buildings. Sticky papers are distributed by pairs in the darkest corners of such buildings, protected from the wind and at uniform distances from the walls (5-8 cm) or under the ceiling (0.5 m). It has been calculated that in the larger settlements of 200-300 farmsteads, it is sufficient to sample ten farms using 100 sticky papers, these papers being exposed for 24 hours only. On this basis, regular standardized captures are used to provide the necessary information about seasonal incidence and species composition of the phlebotamine population.

The special problems of studying *Phlebotomus* populations in association with their natural ground-burrowing hosts, such as the gerbil, is by no means peculiar to the hot arid regions of Russia. Essentially similar problems exist in many other parts of the world and were the subject, for example, of classical studies in the Sudan in the 'forties. A more recent

example is provided by work in California on the three known species of
Phlebotomus, all of which use burrows of the large ground squirrel (*Citellus
beechyi*) as breeding and diurnal resting sites (Aitken and co-workers, 1975).
In contrast to the Russian approach to this kind of situation, the American
workers in the Sacramento valley employed several study and collecting techniques. In addition to oiled or adhesive-type traps, they used ground or
burrow traps, light traps of various design, animal-baited traps, aspirators
and cigarette smoke. As each method was found to have certain limitations,
the practice adopted was to use a combination of these. Of particular interest
were the burrow traps placed over small and large rodent burrows and
designed to trap flies emerging from the burrows by means of a one-way
funnel. Seasonal densities were determined by using a standard number of
adhesive traps (3), burrow traps (10) and small portable light traps (2)
operated twice a month from 16.00 hours one day until 08.00 the following
day.

A great deal of work has been done on *Phlebotomus* in several other
European countries where leishmaniasis is endemic, and where the natural
hosts are either wild animals such as the fox or domestic animals such as
the dog. Perhaps the most instructive example in the present context, and
one which also provides a most interesting comparison with the Russian work
described above, is provided by the oustanding investigations carried out in
the south of France since 1960, centred in Montpellier (Rioux and Golvan,
1969; Rioux and co-workers, 1979). Cutaneous leishmaniasis exists in
France only in the southern part of the mediterranean coast between Spain
and Italy in the region known as the 'Midi'. During the last twenty years
a team of workers has investigated all aspects of the ecology and epidemiology of this disease, and particularly the behaviour patterns of *Phlebotomus*
in relation to man and the natural host of the *Leishmania* parasite, the dog.
Eight different sampling methods have been examined, and valuable
information obtained about their advantages and limitations, as follows:
(i) daily captures in natural or artificial crevices by means of suction tube;
(ii) nightly captures on walls by the same method; (iii) nightly captures on
human bait; (iv) captures in a baited mosquito net; (v) sticky traps;
(vi) light traps; (vii) CO_2 traps, and (viii) New Jersey mosquito traps
(Rioux and Golvan, 1969).
Method (i), little used previously in Europe, gave very good results.
It had the advantage of pinpointing daytime resting sites, and being non-
selective with regard to anthropophilic and non-anthropophilic species. In
addition, this method collects males as well as females. This method, which
was always productive, occasionally produced very high catches in a short
period, as exemplified by the capture of 103 *Phlebotomus* by four collectors
in 15 minutes in one locality. The ordinary orally-operated suction tube
could be supplemented by one using a battery-operated fan in order to avoid
the risk of inhaling harmful organisms or spores in the process of inhal-
ation.
Method (ii) also gave very good results in habitations. It was found
to be more selective with regard to species, and collected few males. In
method (iii) there is a significant point in technique in that collection was
carried out in complete darkness. The operator, covered in a sleeping bag,
had only his face and hands exposed. Approaching flies could either be
heard in flight or felt as they landed, at which stage they were quickly
sucked into the catching tube moved close to them, without the need to
switch on any distracting light. Only anthropophilic species were caught by
this method, predominantly *P. ariasa*.
The technique of method (iv) is based on the baited net catch used in
mosquito studies (page 64), with the bait in this case being dog or fox.
Method (v) is the sticky trap technique. This has long been used by
the Russian workers, as well as in studies on *Phlebotomus* in Algeria,

Tunisia and Roumania.   Papers, impregnated with castor oil and suspended
or mounted on a frame, were inserted into natural crevices, and were considered to be the most effective non-selective method of sampling.   In the
period from June to September 1964, for example, over 20,000 *Phlebotomus*
were captured in this way by the Montpellier team.   The breakdown of the
catch was as follows:

P. ariasi	4714 males	238 females
P. perniciosus	173 males	5 females
Sergentomyia minuta	5913 males	9191 females

This does reveal, however, that there is selectivity with regard to proportion
of sexes.   in the case of *P. ariasi* and *P. perniciosus* males make up the
bulk of the catch (95% and 97% respectively), while in the case of *S. minuta*
females dominate and the male catch is only 37%
      In method (vi), techniques were developed in accordance with the
observation that light itself does not trap *Phlebotomus*, and that any design
of light-based trap must not only be attractive but must also trap or retain
the flies.   The New Jersey type of light trap incorporating a suction fan was
not found to be effective for *Phlebotomus* because the air movement on the
fringe of the fan zone actually repelled attracted flies before they came
within its effective suction range.   In order to overcome these difficulties the
French workers developed a very ingenious method which appears to have
been little used elsewhere — namely, the use of a low-intensity light source
(as from a pocket torch) installed behind a semi-transparent suspended sticky
paper.   This illuminated oiled-paper trap can take two forms (Figures 9.7,
9.8), within a recess or natural cavity in a wall, or alternatively installed
along the surface of a wall in the form of a 'garland' of illuminated papers.
The illuminated paper trap caught about ten times more flies than non-
illuminated paper, with *Phlebotomus ariasi* the dominant species.
      A further refinement was to standardize the light intensity by con-
necting all pocket torch bulbs to a 6-volt accumulator, which can maintain
constant illumination for many hours at night, and can also be set to switch
on or off automatically.
      In method (vii), based on CO_2, an arrangement of traps was set up to
compare the following combinations: light alone, light plus CO_2, CO_2 alone,
and a control with no light and no CO_2.   Each trap was surmounted by oil-
impregnated paper (Figure 9.9).   The results showed that CO_2 alone is more
attractive than the control, but less attractive than light alone (Table 9.6).
Most attractive of all is the combination of light and CO_2.
      Of all the various sampling methods, the use of oil-impregnated papers
provided the most information about seasonal abundance and about nightly

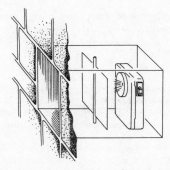

Fig. 9.7   Oiled-paper trap supplemented by low-intensity light source
installed in wall cavity. (After Rioux and Golvan, 1969).

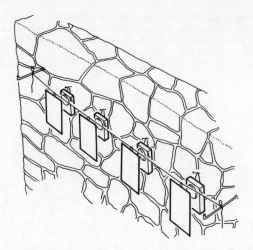

Fig. 9.8   Light traps composed of a 'garland' of illuminated
oiled-paper.   (After Rioux and Golvan, 1969).

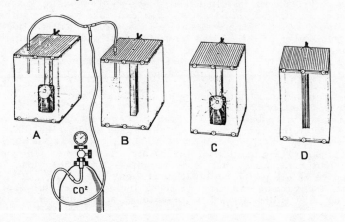

Fig. 9.9   Arrangement for comparing degree of attraction of oiled-paper trap (D), supplemented by CO_2 (B),
by light (C), or both (A).   (After Rioux and Golvan,
1969).

fluctuations in activity (Figures 9.10, 9.11).   These results show that in
France all three dominant species have a single seasonal peak (Figure 9.11).
In contrast, in Tunisia, with a longer summer season, *P. perniciosus* has
two distinct peaks, one at the end of spring and the other in the autumn
(Figure 9.12).

Table 9.6 Captures of *Phlebotomus* in oiled-paper traps supplemented by
CO₂, by light, or by both. (After Rioux and Golvan, 1969).

	CO₂ + light	CO₂ alone	Light alone	Control
P. ariasi				
Male	44	32	31	17
Female	310	60	139	26
S. minuta				
Male	1	0	31	0
Female	2	4	1	3

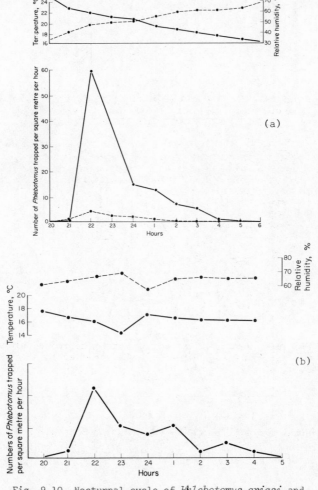

Fig. 9.10 Nocturnal cycle of *Phlebotomus ariasi* and
Sergentomyia minuta based on oiled-paper traps.
(a) Number of both species taken on a warm night.
(b) Number of *P. ariasi* recorded on a relatively
cold and humid night.(After Rioux & Golvan, 1969).
(*P. ariasi* ———   *S. minuta* --------)

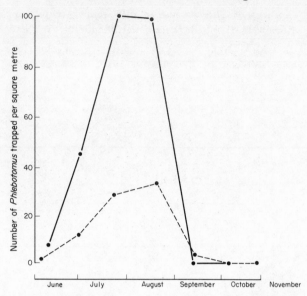

Fig. 9.11 Annual fluctuations in the populations of
Sergentomyia minuta (————) and *Phlebotomus
perniciosus* (--------) in the south of France,
based on oiled-paper trapping data.(After Rioux &
Golvan, 1969).

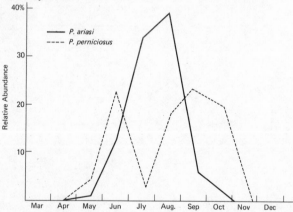

Fig. 9.12 Seasonal abundance of two phlebotamine
vectors of leishmaniasis, i.e. *P. perniciosus* in Tunisia
with two peaks in the year, and *P. ariasi* in France
with a single peak.   Data based on routine trapping
on oiled-adhesive paper.   (After Rioux and Golvan,
1969).

    In more recent studies by a joint Anglo-French team, considerable
attention has been paid to the possibility of mark-release-recapture of adult
Phlebotomus by means of six different colours of fluorescent powder (Rioux
and co-workers, 1979).   Marked flies were found to retain the powder effect-

ively, and could be detected by ultraviolet lamp up to a distance of 6 m.
For recovery of marked flies, captures were made at night by a CDC light
trap, using man or animal bait, and also by collections on walls and other
resting surfaces.   Of particular interest was the observation that with the
marked flies given a blood meal before release, the recapture rate was much
higher than with those released unfed.

References

Chapter 1: Introduction

Huyton, P.M. and Brady, J. (1975) Some effects of light and shade on the feeding and resting behaviour of tsetse flies, *Glossina morsitans* Westwood. J. Ent. (A) 50(1), 23-30.

Jones, M.R.D. and Gubbins, S.J. (1974) Circadian flight activity in four sibling species of the *Anopheles gambiae* complex. Bull Ent. Res. 64, 241-246.

Jones, M.R.D., Hill, M. and Hope, A.M. (1967) The circadian flight activity of the mosquito *Anopheles gambiae*; phase setting by the light regime. J. Exp. Biol. 47, 503-511.

Chapters 2-4: Mosquitoes

Acuff, V.R. (1976) Trap biases influencing mosquito collecting. Mosq. News 36(2), 173-176.

Akiyama, J. (1973) Interpretation of the results of baited trap net collections. J. Trop. Med. Hyg. 76(ii), 283-284.

Bidlingmayer, W.L. (1967) A comparison of trapping methods for adult mosquitoes; species response and environmental influence. J. Med. Ent. 4(2), 200-220.

Bidlingmayer, W.L. (1971) Mosquito flight paths in relation to the environment. I. Illumination levels, orientation and resting areas. Ann. End. Soc. Amer. 64, 1121-1131.

Bidlingmayer, W.L. (1974) The influence of environmental factors and physiological stage on flight patterns of mosquitoes taken in a vehicle aspirator and truck, suction, and New Jersey traps. J. Med. Ent. 11 (2), 119-146.

Bidlingmayer, W.L. (1975) Mosquito flight paths in relation to the environment. Effect of vertical and horizontal visual barriers. Ann. Ent. Soc. Amer. 68(1), 51-57.

Bidlingmayer, W.L., Franklin, B.P., Jennings, A.M. and Cody, E.F. (1974) Mosquito flight paths in relation to the environment. Influence of blood meals, ovarian stage and parity. Ann. Ent. Soc. Amer. 67(6), 919-927.

Bidlingmayer, W.L. and Hem, D.G. (1979) Mosquito (Diptera: Culicidae) flight behaviour near conspicuous objects. Bull Ent. Res. 69, 691-700.

208   Behaviour Patterns of Blood-sucking Flies

Boreham, P.F.L., Chandler, J.A. and Jolly, J. (1978)   The incidence of mos-
    quitoes feeding on mothers and babies at Kisumu, Kenya.   J. Trop. Med
    Hyg. 81, 63–67.
Boreham, P.F.L. and Lenahan, J.K. (1976)   Methods for detecting multiple
    blood-meals in mosquitoes (Diptera: culicidae).   Bull Ent. Res. 66, 671–
    679.
Bowden, J. and Morris, M.G. (1975)   The influence of moonlight on catches of
    insects in light traps in Africa, III. The effective radius of a mercury-
    vapour light trap and the analysis of catches using effective radius.
    Bull. Ent. Res. 65, 303–348.
Bruce-Chwatt, L.J., Garret Jones, G. and Weitz, B. (1966)   Yen years study
    of host selection by anopheline mosquitoes.   Bull. WHO, 35, 405–439.
Carnevale, P., Frezil, J.L., Bosseno, M.F., Le Pont, F. and Lancien, J.
    (1978)   Etude de l'agressivite d'Anopheles gambiae A en function de
    l'age et du sexe des sujets humains.   Bull. World Health Organization,
    56, 147–154.
Chandler, J.A., Highton, R.B. and Boreham, P.F.L. (1976)   Studies on some
    ornithophilic mosquitoes (Diptera: Culicidae) of the Kano Plain, Kenya.
    Bull. Ent. Res. 66, 133–143.
Chandler, J.A., Highton, R.B. and Hill, M.N. (1975)   Mosquitoes of the Kano
    Plain, Kenya.   Part I. Results of indoor collections in irrigated and
    non-irrigated areas using human bait and light traps.   J. Med. Ent.
    12, 504–510.
Chandler, J.A., Parson, J., Boreham, P.F.L. and Gill, G.S. (1977)   Seasonal
    variations in the proportions of mosquitoes feeding on mammals and
    birds at a heronry in western Kenya.   J. Med. Ent. 14(2), 233–240.
Chow, C.Y., Gratz, N.G., Tonn, R.J., Self, L.S. and Pant, C. (1977)   The
    control of Aedes aegypti-borne epidemics.   WHO/VBC/77.660.   WHO Docu-
    mentary Series, Geneva.   16 pp.
Corbet, P.S. and Smith, S.N. (1974)   Diel periodicities of landing of nulli-
    parous and parous Aedes aegypti at Dar es Salaam, Tanzania.   Bull.
    Ent. Res. 64, 111–121.
Davies, J.B. (1978)   Attraction of Culex portesi Senevet & Abonnenc and
    Culex taeniopus Dyar & Knab (Diptera: Culicidae) to 20 animal species
    exposed in a Trinidad forest.   I. Baits ranked by numbers of mosqui-
    toes caught and engorged.   Bull. Ent. Res. 68, 707–719.
Edman, J.D. (1974)   Host-feeding patterns of Florida mosquitoes.   III. Culex
    (Culex) and Culex (Neoculex).   J. Med. Ent. 11(1), 95–104.
Edman, J.D. (1979a)   Host feeding patterns of Florida mosquitoes (Diptera:
    Culicidae).   VI. Culex (Melanoconion).   J. Med. Ent. 15(5-6), 521–525.
Edman, J.D. (1979b)   Orientation of some Florida mosquitoes (Diptera: Culi-
    cidae) toward small vertebrates and carbon dioxide in the field.   J.
    Med. Ent. 15(3), 292–296.
Edman, J.D., Webber, L.A. and Kale, H.W. (1972)   Effect of mosquito density
    on the interrelationship of host behaviour and mosquito feeding success.
    Amer. J. Trop. Med. Hyg. 21(4), 487–491.
Elliott, R. and De Zulueta, J.J. (1975)   Ethological resistance in malaria
    vectors: behavioural responses to intradomestic residual insecticides.
    WHO/VBC/75.569.   WHO Documentary Series, Geneva.   15 pp.
Fontaine, R.E. (1978)   House spraying with residual insecticides with special
    reference to malaria control.   WHO/VBC/78.704.   World Health Organiz-
    ation Documentary Series, Geneva.
Freyvogel, T.A. (1961)   Ein Beitrag zu den Problem um die Blutmahlheit von
    Stechmucken.   Acto Tropica 18, 201–251.
Gillett, J.D. (1979)   Out for blood: flight observations up–wind in the
    absence of visual clues.   Mosq. News, 39(2), 221–229.
Gillies, M.T. (1967)   Experiments on host selection in the Anopheles gambiae
    complex.   Ann. Trop. Med. Parasit. 61, 68–75.

Gillies, M.T. (1969)   The ramp trap, an unbaited device for flight studies of
    mosquitoes. Mosq. News, 29(2), 189–193.
Gillies, M.T., Jones, M.D.R. and Wilkes, T.J. (1978)   Evaluation of a new
    technique for recording the direction of flight of mosquitoes (Diptera:
    Culicidae) in the field.   Bull. Ent. Res. 68, 145–152.
Gillies, M.T. and Wilkes, T.j. (1970)   The range of attraction of single baits
    for some West African mosquitoes.   Bull. Ent. Res. 60(2), 225–235.
Gillies, M.T. and Wilkes, T.j. (1972)   The range of attraction of animal baits
    and carbon dioxide for mosquitoes.   Studies in a freshwater area in
    West Africa.   Bull. Ent. Res. 61(3), 389–404.
Gillies, M.T. and Wilkes, T.j. (1974a)   Evidence of downwind flights by host-
    seeking mosquitoes.   Nature 252, 388–389.
Gillies, M.T. and Wilkes, T.J. (1974b)   The range of attraction of birds as
    baits for some West African mosquitoes.   Bull. Ent. Res. 63, 573–581.
Gillies, M.T. and Wilkes, T.J. (1975)   Long range orientation of *Mansonia*
    males to animal hosts.   Mosq. News, 35(2), 226–227.
Hayes, J. (1975)   Seasonal changes in population structure of *Culex pipiens
    quinquefasciatus* Say (Diptera: Culicidae):   study of an isolated popul-
    ation.   J. Med. Ent. 12, 167–178.
Hayes, R.O., Tempelis, C.H., Hess, A.D. and Reeves, W.C. (1973)   Mosquito
    host preference studies in Hale County, Texas.   Amer. J. Trop. Med.
    Hyg. 22(2), 270–277.
Hess, A.D., Hayes, R.O. and Tempelis, C.H. (1968)   The use of the forage
    ratio technique in mosquito host preference studies.   Mosq. News,
    28(3), 386–389.
Highton, R.B., Bryan, J.H., Boreham, P.F.L. and Chandler, J.A. (1979)
    Studies on the sibling species *Anopheles gambiae* Giles and *Anopheles
    arabiensis* Patton (Diptera: Culicidae) in the Kisumu area, Kenya.
    Bull. Ent. Res. 69, 43–53.
Jones, M.R.D. and Gubbins, S.J. (1974)   Circadian flight activity in four
    sibling species of the *Anopheles gambiae* complex.   Bull. Ent. Res. 64,
    241–246.
jones, M.R.D., Hill, M. and Hope, A.M. (1967)   The circadian flight activity
    of the mosquito *Anopheles gambiae*;   phase setting by the light regime.
    J. Expt. Biol. 47, 503–511.
Jupp, P.G. (1978)   A trap to collect mosquitoes attracted to monkeys and
    baboons.   Mosq. News, 38(2), 288–289.
Jupp, P.G. and McIntosh, B.M. (1967)   Ecological studies on Sindbis and West
    Niles viruses in South Africa.   II. Mosquito bionomics.   S. Afr. J. Med.
    Sci. 32, 15–33.
Jupp, P.G. and McIntosh, B.M. (1970)   Quantitative experiments on the vector
    capability of *Culex (Culex) pipiens fatigans* Wiedmann with West Nile
    and Sindbis viruses.   J. Med. Ent. 7(3), 353–356.
Jurjevskis, I. and Stiles, A.R. (1979)   Summary of stage IV experimental hut
    tests of new insecticides against adult *Anopheles* mosquitoes, 1960–1978.
    WHO/VBC/79.709.   WHO Documentary Series, Geneva.
Kay, B.H., Boreham, P.F.L. and Edman, J.D. (1979)   Application of the
    'feeding index' concept to studies of mosquito host–feeding patterns.
    Mosq. News, 39(1), 68–72.
Khan, A.A., Maibach, H.I. and Strauss, W.G. (1971)   A quantitative study of
    variation in mosquito response and host attractiveness.   J. Med. Ent.
    8(1), 41–43.
Klowden, M.J. and Lea, A.O. (1979)   Effect of defensive host behaviour on
    the blood meal size and feeding success of natural populations of mos-
    quitoes (Diptera: Culicidae).   J. Med. Ent. 15(5–6), 514–517.
Lassen, K., Su-Yung Liu, Lizarzaburu, C. and Rios, R. (1972)   Preliminary
    report on the effect of selective application of propoxur on indoor
    surfaces in El Salvador.   Amer. J. Trop. Med. Hyg. 21(5), 813–818.

Magnarelli, L.A. (1975) Relative abundance and parity of mosquitoes collected in dry-ice baited and unbaited CDC miniature light traps. Mosq. News, 35(3), 350–353.

McCrae, A.W.R. (1972) Age composition of man-biting *Aedes simpsoni* in Bwamba county, Uganda. J. Med. Ent. 9(6), 545–550.

McCrae, A.W.R., Boreham, P.F.L. and Ssenkubuge, Y. (1976) The behaviour ecology of host selection in *Anopheles implexus* Theobald (Diptera: Culicidae). Bull. Ent. Res. 66, 587–631.

McDonald, P.T. (1977a) Population characteristics of domestic *Aedes aegypti* (Diptera: Culicidae) in villages on the Kenya coast. I. Adult survivorship and population size. J. Med. Ent. 14(1), 42–48.

McDonald, P.T. (1977b) Population characteristics of domestic *Aedes aegypti* (Diptera: Culicidae) in villages on the Kenya coast. II. Dispersal within and between villages. J. Med. Ent. 14(1), 49–53.

McIntosh, B.M., Jupp, P.G. and de Sousa, J. (1972) Mosquitoes feeding at two horizontal levels in gallery forest in Natal, South Africa, with reference to possible vectors of chikungunya virus. J. Ent. Soc. S. Afr. 35(1), 81–90.

McIntosh, B.M., Jupp, P.G. and dos Santos, L. (1977) Rural epidemics of Chikungunya in South Africa with involvement of *Aedes (Diceromyia) furcifer* Edwards, and baboons. S. Afr. J. Sci. 73, 267–269.

Muirhead-Thomson, R.C. (1951) The distribution of anopheline mosquito bites among diferent age groups; a new factor in malaria epidemiology. Brit. Med. J., May 19, 1114–1117.

Muirhead-Thomson, R.C. (1958) A pit shelter for sampling outdoor mosquito populations. Bull, WHO. 19, 1116–1118.

Muirhead-Thomson, R.C. (1968) Ecology of Insect Vector Populations. Academic Press, London. 174 pp.

Nelson, M.J., Self, L.S., Pant, C.P. and Usman, S. (1978) Diurnal periodicity of attraction to human bait of *Aedes aegypti* (Diptera: Culicidae) in Jakarta, Indonesia. J. Med. Ent. 14(5), 504–510.

Nelson, R.L., Tempelis, C.H., Reeves, W.C. and Milby, M.M. (1976) Relation of mosquito density to bird:mammal feeding ratios of *Culex tarsalis* in stable traps. Amer. J. Trop. Med. Hyg. 25(4), 644–654.

Pant, C.P. (1979) Vectors of Japanese Encephalitis and the bionomics. WHO/VBC/79.732. WHO Documentary Series, Geneva.

Pant, C.P. and Yasuno, M. (1973) Field studies on the gonotrophic cycle of *Aedes aegypti* in Bangkok, Thailand, J. Med. Ent. 10(2), 219–223.

Pant, C.P. (and others) (1974) A large scale field trial of ultra low volume fenitrothion applied by a portable mist blower for the control of *Aedes aegypti*. Bull. WHO, 51, 409–415.

Paterson, H.E., Brondsen, P., Levitt, J. and Worth, C.B. (1964) Some culicine mosquitoes (Diptera: Culicidae) at Ndumu, Republic of South Africa. Medical Proceedings, 10(9), 188–192.

Port, G.R., Boreham, P.F.L. and Bryan, J.H. (1980) The relationship of host size to feeding by mosquitoes of the *Anopheles gambiae* Giles complex (Diptera: Culicidae). Bull. Ent. Res. 70, 133–144.

Reisen, W.K. and Aalamkham, M. (1978) Biting rhythms of some Pakistan mosquitoes (Diptera: Culicidae). Bull. Ent. Res. 68, 313–330.

Reuben, R. (1971a) Studies of the mosquitoes of North Arcot district, Madras State, India. Part I. Seasonal densities. J. Med. Ent. 8(2), 119–126.

Reuben, R. (1971b) Studies on the mosquitoes of North Arcot district, Madras State, India. Part 2. Biting cycles and behaviour on human and bovine baits at two villages. J. Med. Ent. 8(2), 127–134.

Reuben, R. (1971c) Studies on the mosquitoes of North Arcot district, Madras State, India. Part 3. Host preference for pigs, birds and small mammals. J. Med. Ent. 8(3), 258–262.

Reuben, R. (1971d) Studies on the mosquitoes of North Arcot district, Madras State, India. Part 4. Host preferences as shown by precipitin tests.

J. Med. Ent. 8(3), 314-318.

Reuben, R., Yasuno, M., Panicker, K.N. and LaBrecque, G.C. (1973) The estimation of adult populations of *Aedes aegypti* at two localities in Delhi. J. Comm. Dis. 5(3), 154-162.

Rishikesh, N. and Rosen, P. (1976) Hut entry and exit by *Anopheles gambiae* and *Anopheles funestus* in an unsprayed village near Kaduna in northern Nigeria. Parassitologia, 18, 119-124.

Scholl, P.J., de Foliart, G.R. and Nemenyi, P.B. (1979) Vertical distribution of biting activity by *Aedes triseriatus*. Ann. Ent. Soc. Amer. 72(4), 537-539.

Self, L.S. and Pant, C.P. (1968) Parous/nulliparous condition of unfed *Anopheles gambiae* and *Anopheles funestus* captured in exit traps. Mosq. News, 28(1), 62-64.

Service, M.W. (1971a) Flight periodicities and vertical distribution of *Aedes cantans* (Mg), *Aedes geniculatus* (Ol), *Anopheles plumbeus* Steph. and *Culex pipiens* L. (Diptera: Culicidae) in southern England. Bull. Ent. Res. 60, 639-651.

Service, M.W. (1971b) Feeding behaviour and host preferences of British mosquitoes. Bull. Ent. Res. 60, 653-661.

Service, M.W. (1971c) The daytime distribution of mosquitoes resting in vegetation. J. Med. Ent. 8(3), 271-278.

Service, M.W. (1976) Mosquito Ecology: Field Sampling Methods. Applied Science Publishers, London. 583 pp.

Service, M.W. (1977) A critical review of procedures for sampling populations of adult mosquitoes. Bull. Ent. Res. 67, 343-382.

Shalaby, A.M. (1966) Observations on some responses of *Anopheles culicifacies* to DDT in experimental huts in Gujarat State, India. Ann. Ent. Soc. Amer. 59(5), 938-944.

Sheppard, P.M., Macdonald, W.W., Tonn, R.J. and Grab, B. (1969) The dynamics of an adult population of *Aedes aegypti* in relation to dengue haemorrhagic fever in Bangkok. J. Anim. Ecol. 38. 661-702.

Sinsko, M.J. and Craig, G.B. (1979) Dynamics of an isolated population of *Aedes triseriatus* (Diptera: Culicidae). 1. Population size. J. Med. Ent. 15(2), 89-98.

Snow, W.F. (1975) The vertical distribution of flying mosquitoes in West African savanna. Bull, Ent. Res. 65, 269-277.

Snow, W.F. (1976) The direction of flight of mosquitoes near the ground in West African savannah in relation to wind direction, in the presence and absence of bait. Bull. Ent. Res. 65, 555-562.

Snow, W.F. (1979) The vertical distribution of flying mosquitoes (Diptera: Culicidae) near an area of irrigated rice-fields in the Gambia. Bull. Ent. Res. 69, 561-571.

Snow, W.F. and Boreham, P.F.L. (1973) Feeding habits of some West African *Culex* (Diptera: Culicidae). Bull. Ent. Res. 62, 517-526.

Snow, W.F. and Boreham, P.F.L. (1978) The host-feeding patterns of some culicine mosquitoes (Diptera: Culicidae) in the Gambia. Bull. Ent. Res. 68, 695-706.

Spencer, M. (1967) Anopheline attack on mother and infant pairs, Fergusson Island, Papua New Guinea. Med. J. 10, 75.

Tempelis, C.H. (1975) Host-feeding patterns of mosquitoes, with a review of advances in analysis of blood meals by serology. J. Med. Ent. 11(6), 635-653.

Tempelis, C.H., Hayes, R.O., Hess, A.D. and Reeves, W.C. (1979) Blood-feeding habits of four species of mosquito found in Hawaii. Amer. J. Trop. Med. Hyg. 19, 335-341.

Tempelis, C.H. and Reeves, W.C. (1964) Feeding habits of one anopheline and three culicine mosquitoes by the precipitin test. J. Med. Ent. 1(2), 1248-1251.

Tempelis, C.H. and Washino, R.K. (1967) Host-feeding patterns of *Culex tarsalis* in the Sacramento Valley, California, with notes on other species. J. Med. Ent. 4(3), 315–318.

Trpis, M. and Hausermann, W. (1975) Demonstration of the differential domesticity of *Aedes aegypti* in Africa by mark-release-recapture. Bull. Ent. Res. 65, 199–208.

Vavra, R.W., Carestia, R.R., Frommer, R.L. and Gerberg, E.J. (1974) Field evaluation of alternative light sources as mosquito attractants in the Panama Canal zone. Mosq. News, 34(4), 382–384.

Washino, R.K. and Else, J.G. (1972) Identification of blood meals of haemato-phagus arthropods by the haemoglobin crystallization method. Amer. J. Trop. Med. Hyg. 21(1), 120–122.

Washino, R.K. and Tempelis, C.H. (1967) Host-feeding patterns of *Anopheles freeborni* in the Sacramento Valley, California. J. Med. Ent. 4(3), 311–314.

White, G.B. (1974) *Anopheles gambiae* complex and disease transmission in Africa. Trans. Roy. Soc. Trop. Med. Hyg. 68, 278–298.

Wilton, D.P. (1975) Mosquito collections in El Salvador with ultraviolet and CDC miniature light traps with and without dry ice. Mosq. News, 35(4), 522–525.

Wilton, D.P. and Fay, R.W. (1972) Air flow direction and velocity in light trap design. Ent. Exp. et Applic. 15, 377–386.

World Health Organization (1971a) Vector surveillance: developing standard methods. WHO Chronicle, 268–273.

World Health Organization (1971b) Prevention of *Aedes aegypti*-borne diseases in the Americas. WHO Chronicle, 275–279.

World Health Organization (1980) Yellow fever in 1978. WHO Chronicle, 34, 1–34.

Wright, J.W. (1971) The WHO programme for the evaluation and testing of new insecticides. Bull. WHO, 44, 11–22.

Chapter 5:   Tsetse Flies

Boyt, W.P., MacKenzie, P.K.I. and Pilson, R.D. (1978) The relative attrac-tiveness of donkeys, cattle, and sheep and goats to *Glossina morsitans morsitans* Westwood and *G. pallidipes* Austen (Diptera: Glossinidae) in a middle veld area of Rhodesia. Bull. Ent. Res. 68, 497–500.

Bursell, E. (1970) Theoretical aspects of the control of *Glossina morsitans* by game destruction. Zool. Africana 5(1), 135–141.

Bursell, E. (1973) Entomological aspects of the epidemiology of sleeping sickness. Cent. Afr. J. Med. 19(9), 201–204.

Buxton, P.A. (1955) The Natural History of Tsetse Flies. H.K. Lewis, London. 816 pp.

Challier, A. and Laveissiere, C. (1973) Un noveau piege pour la capture des glossines (Glossina: Diptera, Muscidae): description et essais sur le terrain. Cahiers ORSTOM ser Ent. Med. et Parasit., XI(4), 251–262.

Ford, J. (1971) The Role of the Trypanosomiases in African Ecology. Clarendon Press, Oxford. 950 pp.

Ford, J., Maudlin, I. and Humphryes, K.C. (1972) Comparisons between three small collections of *Glossina morsitans morsitans* (Machado) (Dip-tera: glossinidae) from the Kilombero River valley, Tanzania. Part I. Characteristics of flies exhibiting different patterns of behaviour. Acta Tropica 29, 231–249.

Glasgow, J.P. (1963) Distribution and Abundance of Tsetse. Pergamon Press, Oxford, 241 pp.

Hargrove, J.W. (1976) The effect of human presence on the behaviour of tsetse (*Glossina* spp.) (Diptera: Glossinidae) near a stationary ox.

Bull. Ent. Res. 66, 173–178.
Hargrove, J.W. (1977) Some advances in the trapping of tsetse (Glossina spp.) and other flies. Ecol. Ent. 2, 123–137.
Hargrove, J.W. and Vale, G.A. (1978) The effect of host odour concentration on catches of tsetse flies (Glossinidae) and other Diptera in the field. Bull. Ent. Res. 68, 607–612.
Harley, J.M.B. (1965) Activity cycles of Glossina pallidipes Aust., G. palpalis fuscipes Newst. and G. brevipalpis Newst. Bull. Ent. Res. 56, 141–160.
Huyton, P.M. and Brady, J. (1975) Some effects of light and shade on the feeding and resting behaviour of tsetse flies, Glossina morsitans Westwood. J. Ent. (A), 50(1), 23–30.
Koch, K. and Spielberger, U. (1979) Comparison of hand-nets, biconical traps and an electric trap for sampling Glossina palpalis palpalis (Robineau-Desvoidy) and G. tachinoides Westwood (Diptera: Glossinidae) in Nigeria. Bull. Ent. Res. 69, 243–253.
Laveissiere, C., Couret, D. and Challier, A. (1979) Description and design details of a biconical trap used in the control of tsetse flies along the banks of rivers and streams. WHO/VBC/79.746. WHO Documentary Series, Geneva.
Muirhead-Thomson, R.C. (1968) Ecology of Insect Vector Populations. Academic Press, London. 174 pp.
Mulligan, H.W. (Ed.) (1970) The African Trypanosomiases. Allen & UNwin, London. 950 pp.
Phelps, R.J. (1968) A falling cage for sampling tsetse flies (Glossina, Diptera). Rhod. J. Agric. Res. 6, 47–53.
Phelps, R.J. and Vale, G.A. (1978) Studies on populations of Glossina morsitans and G. pallidipes (Diptera: Glossinidae) in Rhodesia. J. Appl. Ecol. 15, 743–760.
Pilson, R.D., Boyt, W.P. and MacKenzie, P.K.I. (1978) The relative attractiveness of cattle, sheep and goats to Glossina morsitans morsitans Westwood and G. pallidipes Austen (Diptera: Glossinidae) in the Zambezi Valley of Rhodesia. Bull. Ent. Res. 68, 489–495.
Pilson, R.D. and Leggate (1962a) A diurnal and seasonal study of the feeding activity of Glossina pallidipes Aust. Bull. Ent. Res. 53, 541–549.
Pilson, R.D. and Leggate, B.M. (1962b) A diurnal and seasonal study of the resting behaviour of Glossina pallidipes Aust. Bull. Ent. Res. 53, 551–562.
Pilson, R.D. and Pilson, B.M. (1967) Behavioural studies of Glossina morsitans Westw. in the field. Bull. Ent. Res. 57, 227–257.
Randolph, S.E. and Rogers, D.J. (1978) Feeding cycles and flight activity in field populations of tsetse (Diptera: Glossinidae). Bull. Ent. Res. 68, 655–671.
Rogers, D.J. (1977) Study of a natural populations of Glossina fuscipes fuscipes Newstead and a model of fly movement. J. Animal Ecol. 46, 309–330.
Rogers, ·D.J. and Randolph, S.E. (1978) A comparison of electric-trap and hand-net catches of Glossina palpalis palpalis (Robineau-Desvoidy) and G. tachinoides Westwood (Diptera: Glossinidae) in the Sudan vegetation zone of northern Nigeria. Bull. Ent. Res. 68, 283–297.
Rogers, D.J. and Smith, D.T. (1977) A new electric trap for tsetse flies. Bull, Ent. Res. 67, 153–159.
Scholtz, E., Spielberger, U. and Ali, J. (1976) The night resting sites of the tsetse fly Glossina palpalis palpalis (Robineau-Desvoidy) (Diptera: Glossinidae) in northern Nigeria. Bull Ent. Res. 66, 443–452.
Southwood, T.R.E. (1978) Ecological Methods. 2nd edition. Chapman & Hall, London. 584 pp.
Vale, G.A. (1969) Mobile attractants for tsetse flies. Arnoldia 33(4), 1–6.

Vale, G.A. (1971)   Artificial refuges for tsetse flies (*Glossina* spp.).   Bull.
    Ent. Res. 61, 331–350.
Vale, G.A. (1974a)   Direct observations on the responses of tsetse flies
    Diptera: Glossinidae) to hosts.   Bull. Ent. Res. 64, 589–594.
Vale, G.A. (1974b)   New field methods for studying the responses of tsetse
    flies (Diptera: Glossinidae) to hosts.   Bull. Ent. Res. 64, 199–208.
Vale, G.A. (1974c)   The responses of tsetse flies (Diptera: Glossinidae) to
    mobile and stationary baits.   Bull. Ent. Res. 64, 545–588.
Vale, G.A. (1977)   Feeding responses of tsetse flies (Diptera: Glossinidae) to
    stationary hosts.   Bull. Ent. Res. 67, 635–649.
Vale, G.A. (1979)   Field responses of tsetse flies (Diptera: Glossinidae) to
    odours of men, lactic acid and carbon dioxide.   Bull. Ent. Res. 69,
    459–467.
Vale, G.A. and Cumming, D.H.M. (1976)   The effects of selective elimination
    of hosts on a population of tsetse flies (*Glossina morsitans morsitans*
    Westwood (Diptera: Glossinidae).   Bull. Ent. Res. 66, 713–729.
Vale, G.A. and Phelps, R.J. (1978)   Sampling problems with tsetse flies
    Diptera: Glossinidae).   J. Appl. Ecol. 15, 715–726.

Chapter 6:   Tabanids

Adkins, T.R., Ezell, W.B., Sheppard, D.C. and Askey, M.M. (1972)   A modi-
    fied canopy trap for collecting Tabanidae (Diptera).   J. Med. Ent.
    9(2), 183–185.
Anderson, J.R., Olkowsky, W. and Hoy, J.B. (1974)   The response of Tabanid
    species to CO_2-baited insect flight traps in Northern California.
    Pan-Pacific Ento. 50(3), 255–268.
Duke, B.O.L. (1955)   Studies on the biting habits of *Chrysops*.   III. The
    effect of groups of persons, stationary and moving, on the biting
    density of *Chrysops silacea* at ground level in the rain forest at
    Kumba, British Cameroons.   Ann. Trop. Med. Parasit. 49, 362–367.
Duke, B.O.L. (1958)   Studies on the biting habits of *Chrysops*.   V. The
    biting cycles and infection rates of *C. silacea*, *C. dimidiata*, *C. langi*
    and *C. centurionis* at canopy level in the rain-forest at Bombe,
    British Cameroons.   Ann. Trop. Med. Parasit. 52, 24–35.
Duke, B.O.L. (1959)   Studies on the biting habits of *Chrysops*.   VI. A com-
    parison of the biting habits, monthly densities and infection rates of
    C. silacea and *C. dimidiata* (Bombe form) in the rain-forest at Kumba,
    Southern Cameroons, UUKA.   Ann. Trop. Med. Parasit. 53, 203–214.
Gordon, R.M., Kershaw, W.E., Crewe, W. and Oldroyd, H. (1950) The problem
    of loaisis in West Africa, with special reference to recent investigations
    at Kumba in the British Cameroons and at Sapele in southern Nigeria.
    Trans. Roy. Soc. Trop. Med. Hyg. 44, 11.
Haddow, A.J. (1952)   Further observations on the biting habits of Tabanidae
    in Uganda.   Bull. Ent. Res. 42, 659–674.
Haddow, A.J. and Corbet, P.S. (1960)   Observations on nocturnal activity in
    some African Tabanidae (Diptera).   Proc. Roy. Ent. Soc. Lond. (A),
    35(1–3), 1–5.
Haddow, A.J., Gillett, J.D., Mahaffy, A.F. and Highton, R.B. (1950)   Observ-
    ations on the biting habits of some Tabanidae in Uganda, with special
    reference to arboreal and nocturnal activity.   Bull. Ent. Res. 41,
    209–221.
Harlan, D.P. and Roberts, R.H. (1976)   Tabanidae: use of a self-marking
    device to determine populations in the Mississippi-Yazoo river delta.
    Env. Ent. 5(2), 210–212.
Harley, J.B.M. (1965)   Seasonal abundance and diurnal variations in activity
    of some Stomoxys and Tabanidae in Uganda.   Bull. Ent. Res. 56, 319.

Krinsky, W.L. (1976) Animal disease agents transmitted by horse flies and deer flies (Diptera: Tabanidae). J. Med. Ent. 13(3), 225–275.

Morris, K.R.S. (1963) A study of African Tabanids made by trapping. Acta Tropica 20(1), 16–34.

Mullens, B.A. and Gerhardt, R.R. (1979) Feeding behaviour of some Tennessee tabanids. Env. Ent. 8, 1047–1051.

Okiwelu, S.N. (1975) Seasonal distribution and variations in diurnal activity of Tabanidae in the Republic of Zambia. Mosq. News, 35, 551–554.

Okiwelu, S.N. (1977) Preliminary observations of Tabanidae at a tsetse picket in the Republic of Zambia. J. Med. Ent. 14(2), 201–203.

Roberts, R.H. (1972) The effectiveness of several types of Malaise traps for the collection of Tabanidae and Culicidae. Mosq. News, 32(4), 542–547.

Roberts, R.H. (1975) Influence of trap screen age on collections of Tabanids in Malaise traps. Mosq. News, 35(4), 538–539.

Roberts, R.H. (1976a) Altitude distribution of Tabanidae as determined by Malaise trap collections. Mosq. News, 36(4), 518–520.

Roberts, R.H. (1976b) The comparative efficiency of six trap types for the collection of Tabanidae (Diptera). Mosq. News, 36(4), 530–535.

Roberts, R.H. (1977) Attractancy of two black decoys and CO_2 to Tabanids. (Diptera: Tabanidae). Mosq. News, 37(2), 169–172.

Tallamy, D.W., Hansens, E.J. and Denno, R.F. (1976) A comparison of Malaise trapping and serial netting for sampling a horsefly and deerfly community. Env. Ent. 5(4), 788–792.

Tashiro, H. and Schwardt, H.H. (1949) Biology of the major species of horse flies in central New York. J. Econ. Ent. 42, 269–272.

Thompson, P.H. (1969) Collecting methods for Tabanidae. Ann. Ent. Soc. Amer. 62(1), 50–57.

Thompson, P.H. and Bregg, E.J. (1974) Structural modifications and performance of the modified animal trap and the modified Manitoba trap for collection of Tabanidae (Diptera). Proc. Ent. Soc. Wash. 76(2), 119–122.

Thompson, P.H. and Pechuman, L.L. (1970) Sampling populations of *Tabanus quinquevittatus* about horses in New Jersey, with notes on the identity and ecology. J. Econ. Ent. 63(1), 151–155.

Vale, G.A. and Phelps, R.J. (1974) Notes on the host–finding behaviour of Tabanidae (Diptera). Arnoldia, Rhodesia, 36(6), 1–6.

Chapter 7:   Blackflies

Alverson, D.R. and Noblet, R. (1976) Response of female black flies to selected meteorological factors. J. Med. Ent. 9, 662–665.

Anderson, J.R. and de Foliart, G.R. (1961) Feeding behaviour and host preferences of some black flies (Diptera: Simuliidae) in Wisconsin. Ann. Ent. Soc. Amer. 54, 716–729.

Anderson, J.R., Trainer, D.O. and de Foliart, G.R. (1962) Natural and experimental transmission of the waterfowl parasite, *Leucocytozoon simondi* M & L, in Wisconsin. Zoonoses Research, 1(9), 155–164.

Baldwin, W.F. and Gross, H.P. (1972) Fluctuations in numbers of adult black flies (Diptera: Simuliidae) in Deep River, Ontario. Canad. Ent. 104, 1465–1470.

Baldwin, W.F., West, A.R. and Gomery, J. (1975) Dispersal patterns of blackflies (Diptera: Simuliidae) tagged with ^{32}P. Canad. Ent. 107(2), 113–118.

Bellec, C. (1976) A new sampling method for adult *Simulium damnosum* Theobald, 1903 (Diptera: Simuliidae). WHO/VBC/76.642. World Health Organization Documentary Series, Geneva.

Bennet, G.F. (1960) On some ornithophilic blood-sucking Diptera in Algonquin Park, Ontario, Canada. Canad. J. Zool. 38, 377–389.

Bennet, G.F. (1963) Use of P^{32} in the study of a population of *Simulium rugglesi* (Diptera: Simuliidae) in Algonquin Park, Ontario. Canad. J. Zool. 41, 831–840.

Bennet, G.F. and Fallis, A.M. (1971) Flight range, longevity, and habitat preference of female *Simulium euryadminiculum* Davies (Diptera: Simuliidae). Canad. J. Zool. 49(9), 1203–1207.

Bennet, G.F., Fallis, A.M. and Campbell, A.G. (1972) The response of *Simulium (Eusimulium) euryadminiculum* Davies (Diptera: Simuliidae) to some olfactory and visual stimuli. Canad. J. Zool. 50, 793–800.

Colbo, M.H. and Moorhouse, D.E. (1974) The survival of the eggs of *Austrosimulium pestilens* Mack. and Mack. (Diptera: Simuliidae). Bull. Ent. Res. 64, 629–632.

Davies, J.B., Le Berre, R., Walsh, J.F. and Cliff, B. (1978) Onchocerciasis and *Simulium* control in the Volta River Basin. Mosq. News, 38, 466–472.

Davies, L. and Roberts, D.M. (1973) A net and a catch-segregating apparatus mounted in a motor vehicle for field studies on flight activity of Simuliidae and other insects. Bull. Ent. Res. 63, 103–112.

Davies, L. and Williams, C.B. (1962) Studies on black flies (Diptera: Simuliidae) taken in a light trap in Scotland. I. Seasonal distribution, sex ratio and internal condition of catches. Trans. Roy. Ent. Soc. Lond. 114, 1–20.

DeFoliart, G.R. and Rao, R.M. (1965) The ornithophilic black fly *Simulium meridionale* Riley (Diptera: Simuliidae) feeding on man during autumn. J. Med. Ent. 2, 84–85.

Disney, R.H.L. (1968) The timing of adult eclosion in blackflies (Diptera: (Simuliidae) in West Cameroon. Bull. Ent. Res. 59, 485–503.

Disney, R.H.L. (1970) The timing of the first blood meal in *Simulium damnosum* Theobald. Ann. Trop. Med. Parasit. 64(1), 123–128.

Disney, R.H.L. (1972) Observations on chicken-biting blackflies in Cameroon with a discussion of parous rates of *Simulium damnosum*. Ann. Trop. Med. Parasit. 66(1), 149–158.

Disney, R.H.L. (1975) A survey of blackfly populations (Diptera: Simuliidae) in West Cameroon. Ento. Month. Mag. 111, 211–226.

Disney, R.H.L. and Boreham, P.F.L. (1969) Blood gorged resting blackflies in Cameroon and evidence of zoophily in *Simulium damnosum*. Trans. Roy. Soc. Trop. Med. Hyg. 63(2), 286–287.

Duke, B.O.L. (1975a) The differential dispersal of nulliparous and parous *Simulium damnosum*. Zeit. Tropenmed. Parasit. 26, 88–97.

Duke, B.O.L. (1975b) The *Onchocerca volvulus* transmission potentials and associated patterns of onchocerciasis at four Cameroon Sudan-savannah villages. Zeit. Tropenmed. Parasit. 26, 143–154.

Duke, B.O.L., Lewis, D.J. and Moore, P.J. (1966) *Onchocerca-Simulium* complexes. I. Transmission of forest and Sudan-savannah strains of *Onchocerca volvulus*, from Cameroon, by *Simulium damnosum* from various West African bioclimatic zones. Ann. Trop. Med. Parasit. 60(3), 318–336.

El Bashir, S., El Jack, M.H. and El Hadi, H.M. (1976) The diurnal activity of the chicken-biting black fly, *Simulium griseicolle* Becker (Diptera: Simuliidae) in northern Sudan. Bull. Ent. Res. 66, 481–487.

Fallis, A.M. (1964) Feeding and related behaviour of female Simuliidae (Diptera). Expt. Parasit. 15(5), 439–470.

Fallis, A.M., Bennet, G.F., Griggs, G. and Allen, T. (1967) Collecting *Simulium venustum* females in fan traps and on silhouettes with the aid of carbon dioxide. Canad. J. Zool. 45, 1011-1-17.

Fredeen, F.J.H. (1958) Black flies (Diptera: Simuliidae) of the agricultural areas of Manitoba, Saskatchewan and Alberta. Proc. 10th Inter.

Congress Ent. 3, 819–823.

Fredeen, F.J.H. (1961)   A trap for studying the attacking behaviour of black flies, *Simulium arcticum* Mall.   Canad. Ent. 93, 73–78.

Fredeen, F.J.H. (1963)   Oviposition in relation to the accumulation of bloodthirsty black flies *Simulium (Gnus) arcticum* Mall. (Diptera) prior to a damaging outbreak.   Nature, 200, 1024.

Fredeen, F.J.H. (1969)   Outbreaks of the black fly *Simulium arcticum* Malloch in Alberta.   Quaest. Ent. 5, 341–372.

Fredeen, F.J.H. (1977)   A review of the economic importance of black flies (Simuliidae) in Canada.   Quaest. Ent. 13, 219–229.

Fredeen, F.J.H., Spinks, J.W.T., Anderson, J.R., Arnason, A.P. and Rempel, J.G. (1953)   Mass tagging of black flies (Diptera: Simuliidae) with radiophosphorus.   Canad. J. Zool. 31, 1–15.

Garms, R. (1973)   Quantitative studies on the transmission of *Onchocerca volvulus* by *Simulium damnosum* in the Bong Range, Liberia.   Zeit. Tropenmed. Parsit. 24, 358–372.

Garms, R. (1978)   Use of morphological characters in the study of *Simulium damnosum* s.l. populations in West Africa.   Tropenmed. Parasit. 29, 483–491.

Garms, R. and Vajime, C.G. (1975)   On the ecology and distribution of the species of the *Simulium damnosum* complex in different bioclimatic zones of Liberia and Guinea.   Zeit. Tropenmed. Parasit. 9(5), 269–276.

Garms, R. and Voelker, J. (1969)   Unknown filarial larvae and zoophily in *Simulium damnosum* in Liberia.   Trans. Roy. Soc. Trop. Med. Hyg. 63(5), 676–677.

Guttman, D. (1972)   The biting activity of black flies (Diptera: Simuliidae) in three types of habitats in Western Colombia.   J. Med. Ent. 9(5), 269–276.

Hunter, D.M. and Moorhouse, D.E. (1976)   Comparative bionomics of adult *Austrosimulium pestilens* Mackerras & Mackerras and *A. bancrofti* Taylor (Diptera: Simuliidae).   Bull. Ent. Res. 66, 453–467.

Le Berre, R. (1966)   Contributions a l'etude biologique et ecologique de *Simulium damnosum* Theobald, 1903 (Diptera: Simuliidae).   Mem. Off. Rech. Sci. tech, Outre–mer No. 17.   204 pp.

Lewis, D.J. and Duke, B.O.L. (1966)   *Onchocerca–Simulium* complexes. II. Variation in West African *Simulium damnosum*.   Ann. Trop. Med. Parasit. 60(3), 337–346.

Marr, J.D.M. (1971)   Observations on resting *Simulium damnosum* (Theobald) at a dam site in northern Ghana.   WHO/Ohcho/71.85 and WHO/VBC/71.298. World Health Organization Documentary Series, Geneva.

Moore, H.S. and Noblet, R. (1974)   Flight range of *Simulium slossonae*, the primary vector of *Leucocytozoon smithi* of Turkeys in South Carolina. Env. Ent. 3(3), 365–369.

Muirhead–Thomson, R.C. (1956)   Communal oviposition in *Simulium damnosum* Theobald (Diptera: Simuliidae).   Nature, 178, 1297–1299.

Peterson, D.G. and Wolfe, L.S. (1958)   The biology and control of black flies (Diptera: Simuliidae) in Canada.   Proc. 10th Inter. Congress Ent. 3, 551–564.

Quillevere, D., Pendriez, B., Sechan, Y. and Philippon, B. (1977)   Etude du complexe *Simulium damnosum* en Afrique de l'ouest. VII. Etude de la bioecologie et du pouvoir vecteur des femelles de *S. sanctipauli*, *S. soubrense* et *S. yahense* en Cote d'Ivoire.   Cahiers ORSTOM, Entomologie Medicale et Parasit. 15(4), 301–329.   Bouake, Ivory Coast.

Quillevere, D., Philippon, B., Sechan, Y. and Pendriez, B. (1978)   Etude du complexe *Simulium damnosum* en Afrique de l'ouest. VIII. Etude de la bioecologie et du pouvoir vecteur des femelles de savane.   Comparison avec les femelles de foret.   Cahiers ORSTOM, Ent. Med. et Parasit. 16(2), 151–164.

Service, M.W. (1977) Methods for sampling adult Simuliidae, with special reference to the *Simulium damnosum* complex. Centre for Overseas Pest Research, Tropical Pest Bulletin, No. 5, 45 pp.

Service, M.W. (1979) Light trap collections of ovipositing *Simulium damnosum* in Ghana. Ann. Trop. Med. Parasit. 73(5), 487–490.

Thompson, B., Walsh, J.F. and Walsh, B. (1972) A marking and recapture experiment on *Simulium damnosum* and bionomic observations. WHO/Oncho/72.98. World Health Organization Documentary Series, Geneva.

Thompson, B.H. (1976a) Studies on the flight range and dispersal of *Simulium damnosum* (Diptera: Simuliidae) in the rain forest of Cameroon. Ann. Trop. Med. Parasit. 70(3), 343–353.

Thompson, B.H. (1976b) Studies on the attraction of *Simulium damnosum* s.l. (Diptera: Simuliidae) to its hosts. I. The relative importance of sight, exhaled breath, and smell. Zeit. Tropenmed. Parasit. 27, 455–473.

Thompson, B.H. (1976c) Studies on the attraction of *Simulium damnosum* s.l. (Diptera: Simuliidae) to its hosts. II. The nature of substances on the human skin responsible for attractant olfactory stimuli. Zeit. Trop. Parasit. 27, 83–90.

Thompson, B.H. (1976d) The intervals between the blood-meals of man-biting *Simulium damnosum* (Diptera: Simuliidae). Ann. Trop. Med. Parasit. 70(3), 329–341

Vajime, C. and Quillevere, D. (1978) The distribution of the *Simulium damnosum* complex in West Africa with particular reference to the Onchocerciasis Control Programme area. Zeit. Tropenmed. Parasit. 29, 473–482.

Walsh, J.F. (1972) Observations on the resting of *Simulium damnosum* in trees near a breeding site in the West African savanna. WHO/Oncho/72-99. World Health Organization Documentary Series, Genevea.

Walsh, J.F. (1978) Light trap studies on *Simulium damnosum* s.l. in Northern Ghana. Tropenmed. Parasit. 29, 492–496.

Walsh, J.F., Davies, J.B., Le Berre, R. and Garms, R. (1978) Standardization of criteria for assessing the effect of *Simulium* control in onchocerciasis control programmes. Trans. Roy. Soc. Trop. Med. Hyg. 72, 675–676.

Wolfe, L.S. and Peterson, D.G. (1960) Diurnal behaviour and biting habits of black flies (Diptera: Simuliidae) in the forests of Quebec. Canad. J. Zool. 38, 489–497.

World Health Organization (1966) WHO Expert Committee on Onchocerciasis. Second Report. Geneva, Technical Rep. Ser. No. 335. 96 pp.

World Health Organization (1977) Onchocerciasis Control Programme in the Volta River Basin area. Scientific, Technical and Advisory Committee. Report of the 6th Meeting, Geneva, November 1977.

World Health Organization (1978) Onchocerciasis Control Programme in the Volta River Basin area. Evaluation Report, Part 1. WHO, Geneva, OCP/78.2.

Chapter 8: Biting Midges

Davies, J.E. and Giglioli, M.E.C. (1977) The breeding sites and seasonal occurrence of *Culicoides furens* in Grand Cayman with notes on the breeding sites of *Culicoides insiginis* (Diptera: Ceratopogonidae). Mosq. News, 37(3), 414–423.

Davies, J.E. and Giglioli, M.E.C. (1979) Some field methods used in Grand Cayman for trapping adult ceratopogonids (Diptera). Mosquito News, 39, 149–153.

Kettle, D.S. (1965) Biting Ceratopogonidae as vectors of human and animal diseases. Acta Tropica, 22(4), 356–362.

Kettle, D.S. (1969a) The biting habits of *Culicoides furens* (Poey) and *C. barbosai* Wirth & Blanton. I. The 24-hour cycle, with a note on differences between collectors. Bull. Ent. Res. 59, 21-31.

Kettle, D.S. (1969b) The biting habits of *Culicoides furens* (Poey) and *C. barbosai* Wirth & Blanton. II. Effect of meteorological conditions. Bull. Ent. Res. 59, 241-258.

Kettle, D.S. (1969c) The ecology and control of blood-sucking Ceratopogonidae. Acta Tropica, 26(3), 235-248.

Kettle, D.S. and Linley, J.R. (1967a) The biting habits of *Leptoconops bequaerti*. I. Methods; standardization of techniques; preference for individuals, limbs and positions. J. App. Ecol. 4, 379-395.

Kettle, D.S. and Linley, J.R. (1967b) The biting habits of *Leptoconops bequaerti*. II. Effect of meteorological conditions on biting activity; 24 hour and season cycles. J. Appl. Ecol. 4, 397-420.

Kettle, D.S. and Linley, J.R. (1969a) The biting habits of some Jamaican *Culicoides*. I. *C. barbosai* Wirth & Blanton. Bull. Ent. Res. 58, 729-753.

Kettle, D.S. and Linley, J.R. (1969b) The biting habits of some Jamaican *Culicoides*. II. *C. furens* (Poey). Bull. Ent. Res. 59, 1-20.

Kline, D.L. and Axtell, R.C. (1976) Salt marsh *Culicoides* (Diptera: Ceratopogonidae) species, seasonal abundance and comparisons of trapping methods. Mosq. News, 36(1), 1-10.

Koch, H.G. and Axtell, R.C. (1979a) Attraction of *Culicoides furens* and *C. hollensis* (Diptera: Ceratopogonidae) to animal hosts in a salt marsh habitat. J. Med. Ent. 15(5-6), 494-499.

Koch, H.G. and Axtell, R.C. (1979b) Correlation of hourly suction trap collections of *Culicoides furens* and *C. hollensis* (Diptera: Ceratopogonidae) with wind, temperature and habitat. J. Med. Ent. 15(5-6), 500-505.

Linley, J.R. and Davies, J.B. (1971) Sandflies and tourism in Florida and the Bahamas and Caribbean area. J. Econ. Ent. 64, 264-278.

Nelson, R.L. and Bellamy, R.E. (1971) Patterns of flight activity of *Culicoides variipennis* (Coquillet) (Diptera: Ceratopogonidae). J. Med. Ent. 8(3), 283-291.

Service, M.W. (1974) Further results of catches of *Culicoides* (Diptera: Ceratopogonidae) and mosquitoes from suction traps. J. Med. Ent. 11(4), 471-479.

Tanner, G.D. and Turner, E.C. (1974) Vertical activities and host preference of several *Culicoides* species in a southwestern Virginia forest. Mosq. News, 34(1), 66.

Tanner, G.D. and Turner, E.C. (1975) Seasonal abundance of *Culicoides* spp. as determined by three trapping methods. J. Med. Ent. 12, 87-91.

Turner, E.C. (1972) An animal-baited trap for the collection of *Culicoides* spp. (Diptera: Ceratopogonidae). Mosq. News, 32(4), 527-530.

Walker, A.R. (1977) Adult lifespan and reproductive status of *Culicoides* species (Diptera: Ceratopogonidae) in Kenya with reference to virus transmission. Bull. Ent. Res. 67, 205-215.

Walker, A.R. and Boreham, P.F.L. (1976) Blood-feeding of *Culicoides* (Diptera: Ceratopogonidae) in Kenya in relation to the epidemiology of blue tongue, an ephemeral fever. Bull. Ent. Res. 66, 181-188.

Chapter 9:   Phlebotomine Sandflies

Aitken, T.H.G., Woodall, J.P., de Andrade, A.H.P., Bensabath, G. and Shope, R.E. (1975) Pacui virus, phlebotamine flies, and small mammals in Brazil: an epidemiological study. Amer. J. Trop. Hyg. 24, 358-368.

Chaniotis, B.N. and Anderson, J.R. (1968) Age structure, population dynamics and vector potential of *Phlebotomus* in northern California. Part II. Field population dynamics and natural flagellate infections in parous females. J. Med. Ent. 5(3), 273-292.

Chaniotis, B.N. and Correa, M.A. (1974) Comparative flying and biting activity of Panamanian phlebotomine sandflies in a mature forest and adjacent open space. J. Med. Ent. 11(1), 115-166.

Chaniotis, B.N., Correa, M.A., Tesh, R.B. and Johnson, K.M. (1971a) Daily and seasonal man-biting activity of phlebotomine sandflies in Panama. J. Med. Ent. 8(4), 415-420.

Chaniotis, B.N., Correa, M.A., Tesh, R.B. and Johnson, K.M. (1974) Horizontal and vertical movements of phlebotomine sandflies in a Panamanian rain forest. J. Med. Ent. 11(3), 369-375.

Chaniotis, B.N., Neely, J.M., Correa, M.A., Tesh, R.B. and Johnson, K.M. (1971b) Natural population dynamics of phlebotomine sandflies in Panama. J. Med. Ent. 8(4), 339-352.

Chaniotis, B.N., Tesh, R.B., Correa, M.A. and Johnson, K.M. (1972) Diurnal resting sites of phlebotomine sandflies in a Panamanian tropical forest. J. Med. Ent. 9, 91-98.

Dergacheva, T.I., Zherikhina, I.I. and Rasnitsyna, N.M. (1979) A method of counting sandflies (Diptera: Phlebotominae). WHO/VBC/79.718. World Health Organization Documentary Series, Geneva.

Disney, R.H.L. (1966) A trap for phlebotomine sandflies attracted to rats. Bull. Ent. Res. 56, 445-451.

Disney, R.H.L. (1968) Observations on a zoonosis : Leishmaniasis in British Honduras. J. Appl. Ecol. 5(1), 1-59.

Gorgas Memorial Laboratory (1977) 48th Annual Report. US Govt. Priting Office. 41 pp.

Killick-Kendrick, R. (1978) Recent advances and outstanding problems in the biology of phlebotomine sandflies. Acta Tropica, 35, 297-313.

Lainson, R. and Shaw, J.J. (1968) Leishmaniasis in Brazil: I. Observations on enzootic rodent leishmaniasis — incrimination of *Lutzomyia flaviscutellata* (Mangabeira) as the vector in the lower Amazonian basin. Trans. Roy. Soc. Trop. Med. Hyg. 62(2), 385-395.

Lainson, R., Shaw, J.J., Ward, R.D. and Fraiha, H. (1973) Leishmaniasis in Brazil: IX. Considerations of the *Leishmania brasiliensis* complex: importance of the sandflies of the genus Psychodopygus (Mangabeira) in the transmission of *L. brasiliensis* in north Brazil. Trans. Roy. Soc. Trop. Med. Hyg. 67(2), 184-196.

Rioux, J.A. and Golvan, Y.J. (1969) Epidemiologie des Leishmanioses dans le sud de la France. Monograph de l'Institut national de la Sante et de la Recherche Medicale, Paris. 223 pp.

Rioux, J.A., Killick-Kendrick, R., Leaney, A.J., Turner, D.P., Bailly, M. and Young, C.J. (1979) Ecologie des leishmanioses dans le sud de la France. 12. Dispersion transversale de *Phlebotomus ariasi* Tonnoir, 1921. Experiences preliminaires. Ann. de Parasit. (Paris), 54 (6), 673-682.

Rutledge, L.C. and Ellenwood, D.A. (1975) Production of phlebotomine sandflies on the open forest floor in Panama; the species complement. Env. Ent. 4(1), 71-77.

Shaw, J.J. and Lainson, R. (1968) Leishmaniasis in Brazil: II. Observations on enzootic rodent leishmaniasis in the lower Amazon region. The feeding habits of the vector, *Luyzomyia flaviscutellata* in reference to man, rodents and other animals. Trans. Roy. Soc. Trop. Med. Hyg. 62(3), 396-405.

Shaw, J.J. and Lainson, R. (1972) Leishmaniasis in Brazil: VI. Observations on the seasonal variations of *Lutzomyia flaviscutellata* in different types of forest and its relationship to enzootic rodent leishmaniasis (*Leishmania mexicana amazonensis*). Trans. Roy. Soc. Trop. Med. Hyg.

66(5), 709–717.
Shaw, J.J., Lainson, R. and Ward, R.D. (1972)   Leishmaniasis in Brazil: VII. Further observations on the feeding habits of *Lutzomyia flaviscutellata* (Mangabeira) with particular reference to its biting habits at different heights. Trans. Roy. Soc. Trop. Med. Hyg. 66(5), 718–723.
Tesh, R.B., Chaniotis, B.N., Aronson, M.D. and Johnson, K.M. (1971) Natural host preferences of Panamanian phlebotomine sandflies as determined by precipitin test.   Amer. J. Trop. Med. Hyg. 20(1), 150–156.
Thatcher, V.E. (1968)   Studies of phlebotomine sandflies using castor oil traps baited with Panamanian animals.   J. Med. Ent. 5(3), 293–297.
Thatcher, V.E. and Hertig, M. (1966)   Field studies on the feeding habits and diurnal shelters of some *Phlebotomus* sandflies (Diptera: Psychodidae) in Panama. Ann. Ent. Soc. Amer. 59(1), 46–52.
Ward, R.D., Shaw, J.J., Lainson, R. and Fraiha, H. (1973)   Leishmaniasis in Brazil: VIII. Observations on the phlebotomine fauna of an area highly endemic for cutaneous leishmaniasis in the Serra dos Carajas, Para State. Trans. Roy. Soc. Trop. Med. Hyg. 67(2), 174–183.
Williams, P. (1965)   Observations on the Phlebotomine sandflies of British Honduras. Ann. Trop. Med. Parasit. 59, 393–404.
Williams, P. (1970)   On the vertical distribution of *Phlebotomus* sandflies (Diptera: Psychodidae) in British Honduras (Belize). Bull. Ent. Res. 59, 637–646.
World Health Organization (1979)   Studies on leishmaniasis vectors/reservoirs and their control in the Old World.   General review and inventory. WHO/VBC/79.749. WHO Documentary Series, Geneva. 88 pp.

Index to Scientific Names of Blood-sucking Flies